T0213617

Electric Power
Distribution Emergency Operation

Electric Power
Distribution Emergency Operation

Chee-Wooi Ten
Michigan Technological University

Yachen Tang
Global Energy Interconnection
Research Institute (GEIRI)
North America

CRC Press
Taylor & Francis Group
Boca Raton London New York

CRC Press is an imprint of the
Taylor & Francis Group, an **informa** business

MATLAB® is a trademark of The MathWorks, Inc. and is used with permission. The MathWorks does not warrant the accuracy of the text or exercises in this book. This book's use or discussion of MATLAB® software or related products does not constitute endorsement or sponsorship by The MathWorks of a particular pedagogical approach or particular use of the MATLAB® software.

CRC Press
Taylor & Francis Group
6000 Broken Sound Parkway NW, Suite 300
Boca Raton, FL 33487-2742

First issued in paperback 2022

© 2019 by Taylor & Francis Group, LLC
CRC Press is an imprint of Taylor & Francis Group, an Informa business

No claim to original U.S. Government works

ISBN-13: 978-1-498-79894-5 (hbk)
ISBN-13: 978-1-03-233888-0 (pbk)
DOI: 10.1201/9780429440830

Library of Congress Cataloging-in-Publication Data

Names: Ten, Chee-Wooi, author. | Tang, Yachen, author.
Title: Electric power : distribution emergency operation / Chee-Wooi Ten and Yachen Tang.
Description: Boca Raton : Taylor & Francis, a CRC title, part of the Taylor & Francis imprint, a member of the Taylor & Francis Group, the academic division of T&F Informa, plc, 2018. | Includes bibliographical references and index.
Identifiers: LCCN 2018027156| ISBN 9781498798945 (hardback : alk. paper) | ISBN 9780429440830 (ebook : alk. paper)
Subjects: LCSH: Electric power distribution--Alternating current. | Emergency power supply.
Classification: LCC TK3141 .T46 2018 | DDC 621.319--dc23
LC record available at https://lccn.loc.gov/2018027156

Visit the Taylor & Francis Web site at
http://www.taylorandfrancis.com

and the CRC Press Web site at
http://www.crcpress.com

Dedicated to Suvie and Shimin

Contents

Preface

The idea for this book was born in 2002 when the first author was a power-system engineer working on the geographic information system (GIS) map, and attempted to model the distribution network topology in the supervisory control and data acquisition (SCADA) system. There was a lack of reference textbooks that could provide a jumpstart for practitioners and novices who had limited training in distribution modeling in the SCADA environment and how it is related to data modeling. Over the past decade, finding a good reference textbook on distribution system modeling with tutorials has been limited. As often happens, the advanced level of graduate distribution network materials is created by the course instructor and depends on the individual focus in distribution systems. The material in this book was implemented in the pilot course, *EE5251–Distribution Engineering II*, which is now a permanent course offered once every two years at Michigan Tech. The authors established the course materials prior to the Fall 2016 semester and received feedback from the graduate students; it was valuable to have different perspectives.

The first author was intrigued by the nature of geographical locations of distribution feeders and laterals whereas most traditional elective power courses mainly focus on the modeling part; this hardly promotes awareness of practically sized distribution network topology of interdependent feeders and the number of elements associated with each feeder. The realization of network topology and complexity only happened when the author was working on a large project where thousands of elements in the electrical distribution network needed to be carefully modeled in the SCADA network.

The authors are extremely grateful for the funding support from the National Science Foundation (NSF) Critical Resilient Interdependent Infrastructure Systems and Processes (CRISP). This three-year project did promote interdisciplinary work and the timing cannot be better, especially because the term *resilience* has been widely promoted in recent years. Fundamentally it is essentially the new definition of reliability in power delivery, a hardened measure of overall system performance that is key. As the electrical grid has been deployed with more intelligent sensors and control technologies, the complexity of distribution networks has gradually evolved toward a more resilient system with reconfigurable possibilities. Resilience for power distribution systems is aimed at a higher expectation of system reliability under natural catastrophes, which requires algorithmic redesign for improving computational efficiency. The fundamentals of graph theory applied to system reconfiguration are the salient points of this book, addressing the energized unbalanced distribution feeders in radial topology.

In this book, the matrices in incidence or adjacency forms provide the visual representation in the ordered format for readers to relate the connected elements. The possibilities of reconfiguration can either be within a feeder, between feeders associated with the same substation, or between feeders from other substations. This will help students to identify from the matrices where troubleshooting may be facilitated in the code based on the expected outcome for possible counts on adjustment and possibilities. Although this book does not provide distributed energy resources, the similar evaluation of other combinations based on distributed elements across primary and secondary elements can be assessed in the same manner.

This book is organized into four major sections. Each section provides focal points of study along with tutorial examples. It is designed for a graduate-level class, which expands the materials beyond W. H. Kersting's distribution system textbook on modeling. All examples are given in MATLAB® script. Each script from each chapter represents the subject of interest for DMS application where it is all connected among others. The seven mini projects from some chapters provide guidance to integrate the modules, with some additions necessary to generate the fault events from a GIS dataset defined by the graduate students. The authors believe this process of integration will give students insight and appreciation of the materials.

Section I of this book is introductory and reviews the state-of-the-art communication architecture of a distribution network. This section provides an overview of how communication can be established between distribution dispatching centers (DDCs), pole-mounted feeder devices, and substation remote terminal units. The deployment of advanced metering infrastructure (AMI) devices improve billing accuracy that enables data exchange with other units. The interdependency between the distribution dispatching center and the billing centers provides not only data reliability of multisite data redundancy, but also data sharing for other applications. This section mainly provides the readers with an understanding of how the interdependent architecture of network communication can be considered for the future establishment of database synchronization. Under extreme emergency situations, such as natural disasters, the operators in the control room may not have all resources available to operate in the best of worst case scenarios with all possibilities.

Section II discusses essential data preparation and modeling for operation. Establishing good-quality topology information leads to higher computational accuracy. The notion of "garbage in, garbage out" is a crucial part of a computerized management system for distribution networks. GIS is a versatile tool that can help planning engineers update the outdated distribution network topology. Each incremental update will be imported to SCADA to reflect the real-world scenarios. This section introduces how GIS datasets can be extracted and then transformed into an incidence or adjacency matrix. This is then further analyzed with the traversing algorithm to determine the energization state of a subsystem. Section III of this book provides a series of chapters that mainly emphasize coordination between the available crew and power outage investigation. The inference of fault segments with immediate isolation of a "healthy" segment within a feeder can be restored from other feeders. This section provides technical details of how the crew can coordinate with the distribution dispatching center. The management applications are used to infer the potential fault segment, and may help the crew search for an exact location and provide a high-level recommendation of switching steps and potential consequences. The trouble call tickets are used to provide additional information for the applications to escalate the outage events and differentiate a single event or possible multiple ones.

A resilient distribution system should not only withstand a fault disturbance but also succeed in more extreme situations, such as natural disasters or intentional foul play by malicious consumers/organizations. Section IV addresses the emerging development on the networked microgrid. This is the subject that addresses extreme circumstances. It is the intent of the authors to introduce graph modeling in the context of energization status and reconfigurability that will be the nexus to their future research exploration.

The authors hope that this book will serve as the distribution engineering II textbook for follow-up graduate courses where the students will gain a more intuitive understanding of the topology in distribution. The geographical topology of interconnected feeders will also be useful for graduate students, whether it is in research or in technical knowledge enrichment in a particular area of interest, which will help them to pursue independent studies.

Course instructors are welcome to request a desk copy when providing contact and course information at `crctextbook@taylorandfrancis.com`. Upon qualifying course adoption, instructors may also request print or e-book desk copies for use in their courses directly. For future updates of MATLAB script files, please visit http://www.crcpress.com/9781498798945.

Chee-Wooi Ten
Michigan Technological University

Yachen Tang
Global Energy Interconnection Research Institute (GEIRI) North America

Acknowledgments

This body of work could not have been accomplished without the tremendous assistance of all my Ph.D. students. They are determined, diligent, and brilliant individuals who made the completion of this book possible. One of them that I must mention here is Yachen Tang, who is also the co-author of this book, and has been extremely devoted to this book project. He has helped to translate my handwritten notes and thoughts (exam questions, slides, etc.) from my EE5251 class. I also would like to acknowledge my first Ph.D. student, Yonghe Guo, who has initiated the concept of graph connectivity that describes outage and energization states in this work.

I am thankful for all the energy from the EE5250 classes over the past eight years. The first distribution engineering class at Michigan Tech was offered, and I have been teaching this course annually since spring 2010. Each time as I taught this course, I felt motivated to work on this book.

Working at Siemens has been my greatest inspiration. I started the job in 2002 and the project has pushed us to master and use the skills on site as we deployed the DMS system. My thanks to Guenter Beissler, Ewald Riedel, Dong Li, and Jens Winkel. Most importantly, I would like to thank Erich Wuergler, Hans-Joachim Diehl, and Hoay Beng Gooi for their mentorship in understanding graphics on SCADA systems, common information modeling, and technical communication. We worked on the publication of GIS importation to SCADA in this work.

I have learned greatly from my doctorate adviser, Chen-Ching Liu. Learning to write persuasively and adapting teaching philosophy from Chen-Ching has been most helpful. Mentors and friends deserve mention here. They are Dennis Morrison, Bruce Wollenberg, Andrew Ginter, and Gerald Sheblé, who have given me very realistic perspectives on power engineering and security to balance theory and practice, particularly in the SCADA/EMS business. I understand that some theoretical work may often dominate or does not bridge well to the real-world scenarios. I truly believe that the innovation of being the best engineering professor needs both.

My thanks to the people who are/were in my life and make me a better person. Their wisdom and encouragement are imprinted in my memory.

Authors

Chee-Wooi Ten is currently an associate professor of electrical and computer Engineering at Michigan Technological University. He received his BSEE and MSEE degrees from Iowa State University, Ames, in 1999 and 2001, respectively, and later received a Ph.D. degree (in 2009) from the University College Dublin (UCD), National University of Ireland.

Professor Ten was a power-application engineer working in project development for EMS/DMS with Siemens Energy Management and Information System (SEMIS) in Singapore from 2002 to 2006 when the idea of this book came about, to help software and power-system engineers working in the distribution dispatching center. His primary research interests are modeling for interdependent critical cyberinfrastructures and SCADA automation applications for a power grid. He is a senior member of the IEEE. He is an active reviewer for IEEE PES transactions and has been a member of the IEEE PES Computer and Analytical Method for Cybersecurity Task Force. Dr. Ten is currently serving as an editor for *IEEE Transactions on Smart Grid*, and for the Elsevier journal *Sustainable Energy, Grids and Networks (SEGAN)*.

Yachen Tang received a Ph.D. degree in electrical and computer engineering at Michigan Technological University in 2018. He earned a BS degree in telecommunication engineering from Jilin University, China, in 2011, and a MS degree in computer engineering from Michigan Technological University, Houghton, in 2014.He is currently with Global Energy Interconnection Research Institute (GEIRI) North America in San Jose.

Dr. Tang was previously an intern with Henan Provincial Computer Research Institute for the project entitled "Testing and Processing System for Network Image of Yellow River," from July 2011 to January 2012. He holds a patent on a program for inputting Chinese characters. Since his doctorate program at Michigan Tech, he has been active in reviewing technical articles for the *IEEE PES* and *IET Generation, Transmission & Distribution* transactions journals. His research interests include power grid cybersecurity, which spans from modeling of anomalies to interference of inconsistent data related to an electrical distribution system. His research interests also include data mining and analysis, and machine learning. In particular, his contribution to this book involves graph theory establishment on the feeders and possible reconfiguration. That fresh view offers a new perception of topology in data structures where the energization states for each element can be affected during power outages for an inference of the root cause.

Abbreviations

A/D	Analog-to-Digital
AMI	Advanced Metering Infrastructure
AMR	Automatic Meter Reading
ASCII	American Standard Code for Information Interchange
BFS	Breadth-First Search
CAD	Computer Aided Design
CAIDI	Customer Average Interruption Duration Index
CB	Circuit Breaker
CIC	Customer Interruption Cost
CID	Customer Interruption Duration
CIM	Common Information Model
CIS	Customer Information System
CRM	Customer Relationship Management
DAG	Directed Acyclic Graph
DBF	Data Base File
DCC	Dispatching Control Center
DDC'	Distribution Dispatching Center
DFS	Depth-First Search
DG	Distribution Generation
DMS	Distribution Management System
DR	Demand Response
DT	Distribution Transformer
EMS	Energy Management Systems

ESRI	Environmental Systems Research Institute
FI	Fault Indicator
FLS	Fault Locating System
FRTU	Feeder Remote Terminal Unit
FTU	Feeder Terminal Unit
GIS	Geographic Information System
GPS	Global Positioning System
HAN	Home Area Network
HES	Head End System
HVAC	Heating, Ventilation, and Air Conditioning
IED	Intelligent Electronic Device
INVC	Investment Cost
IR	Interruption Duration
IT	Information Technology
IVR	Interactive Voice Response
LAN	Local Area Network
MDM	Mobile Device Management
MDMS	Meter Data Management System
NC	Normally Closed
NO	Normally Open
OT	Operational Technology
PLC	Programmable Logic Controller
PMU	Phasor Measurement Unit
RCS	Remote-Controlled Switch
REST	Restoration Time
RF	Radio Frequency
RTU	Remote Terminal Unit
SAIDI	System Average Interruption Duration Index

SAIFI	System Average Interruption Frequency Index
SCADA	Supervisory Control and Data Acquisition
SOE	Sequence-of-Events
SOF	Scheduled Outage Frequency
ST	Substation Transformer
TCM	Trouble Call Management
TCP	Transmission Control Protocol
TCR	Total Cost of Reliability
TID	Total Interruption Duration
WAN	Wide Area Network
XE	Extreme Event

Part I

Network Communication Architecture

1

Computerized Management System

1.1 Introduction

This introductory chapter provides an overview of communication network centers that harness the telemetered information from primary and secondary distribution networks. The supervisory control and data acquisition (SCADA) network archives incoming measurements from the pole-mounted devices in the primary distribution network. The customer-billing center harnesses consumer consumption data from customers' smart meters. The massive deployment of smart meters in many cities would introduce demand response and applications to adjust the peak and non-peak usage with different prices over time. The net metering policies may encourage customers to use more or less energy at different times of the day. The infrastructure of metering distribution networks has become ubiquitous and has enhanced overall system observability, which can derive new applications in the control room to coordinate with the power outages of customers and with work crews as well as to effectively pinpoint the location of the electrical short circuit.

The opportunities and problems in electrical distribution grids are tremendous, particularly in the realm of control and operation [1,2]. Computerized management systems for distribution operation have played a crucial role in reliability improvement [3] as well as in emerging resilience frameworks in reconfiguration [4–6]. Electrical fault current often causes the protective relays to react to the disturbance, resulting in a large area of power outage that may affect thousands of customers. Localizing a faulted segment between all boundary switches can be time-consuming. The information exchange between all entities can help crews to effectively search for a fault location. The reduction of search time as well as restoring power to all customers can be executed in a matter of minutes with the implementation of remote-controlled devices on distribution feeders, which can substantially improve overall system reliability during emergency conditions.

The data exchange introduced in this chapter provides communication network architecture between distribution control centers and customer billing centers. Both entities could have multi-site backup centers for data redundancy in cases where of disruption in the region. A primary operational center can fail over to its backup site in case the communication has been disabled during a critical time.

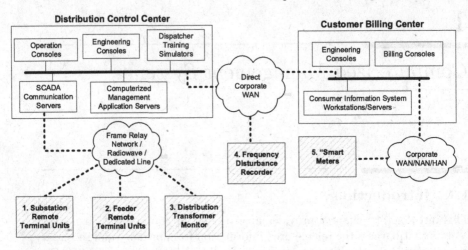

FIGURE 1.1

Communication between distribution control center and customer billing center.

1.2 Distribution Control Center

Distribution control centers are the centralized hubs that are connected with multiple computer servers and client computers for operational purposes [7–11]. This is a critical operation to help distribution companies respond to power outages based on the metering information from the pole-mounted devices deployed on the feeders. Despite the ability to monitor and control the remote devices for possible reconfiguration, the functionalities of SCADA can be limited to optimality based on certain conditions. A distribution management system (DMS) consists of online support applications that utilize the SCADA telemetered measurements as input to compute and provide recommendations in order for operators to make decisions. The modeling of distribution system objects is part of the SCADA network [12,13]. In some cases, the DMS implementation could lead to dynamic pricing schemes among their customers, to increase efficiency of consumption by reducing peak loads from air-conditioning and heating loads within residential buildings [14–17]. The remote-controlled load management is implemented to help a system dispatcher operate based on the best information they receive [18–20].

1.2.1 SCADA Communication Network

The SCADA, simply put as "supervisory control," is a fundamental system for the operation of remote control of a substation's breakers and switch gears [21], or the pole-mounted feeder remote terminal units that established multiple online applications in the system as early as the 1960s [22]. The acronym SCADA abbreviates two major communication pathways, i.e., an ability to control the remote pole-mounted or substation communication devices when necessary as well as to constantly receive the status of the physical system for monitoring purposes. Depending on the bandwidth, sample period, and input data to be analyzed, the investment in increasing communication reliability can be costly [23]. The backbone of SCADA and geographic information system (GIS) networks is shared in the GIS database and maintained separately [24]. In the 1990s, the distribution automation and management system was designed to integrate with existing SCADA systems [25].

The basic function of SCADA is to receive status and analog from sequential events [26,27]. Most importantly, the overall performance of SCADA communication relies heavily on data links between control centers and the remote devices [26]. These telemetered values from the remote sites are often pre-filtered and refined with processing applications in the control center to ensure the quality of datasets [28]. SCADA measurements accumulated from all remote-telemetered devices can provide adequate information to estimate losses in terms of both operation and planning [29].

The deployment of remote terminal units (RTUs) begins at the distribution substation level, which serves as the root node of the associated feeders under a substation [30]. The consideration of deploying a SCADA system is often associated with the costs of building a new distribution substation, and computer automation [31]. The main benefits of SCADA implementation is to receive alarms and measurements that can be used for periodic data analysis, such as the loading condition of a feeder, volt/var control, load growth assessment, and management for each feeder [32–34].

The modern control centers today are networked with large and highly reliable bandwidth [35, 36]. The control variables and measurement points corresponding to each pathway are mapped with physical addresses within its radio frequency intranet. As depicted in Fig. 1.1, the SCADA serves [37] as a platform of data collection from the feeder remote terminal units (FRTUs)/ feeder terminal units (FTUs) or RTUs that is accumulated from primary and secondary distribution networks. Generally, a distribution SCADA network connects to all remote-controlled devices that can be categorized into four types: (1) Substation RTU, (2) FRTU/FTU, (3) distribution transformer monitor, and (4) frequency disturbance units (possible future inclusion for operational purposes). The control area is typically bound to the primary network of a distribution network. The measurement types are categorized into two types: (i) analog and (ii) digital. These measurements are typically associated with a certain geographical location of a feeder where the FRTU/RTU is connected with a disconnecting switch. Control can be executed from a distribution control center to perform a change of status by sending remote commands to an FRTU/RTU. The following control variables are linked to the SCADA addresses with an FRTU/RTU:

- **Open or close a switch**: This is part of the FRTU that enables the remote-controlled capability to open or to close an associated switch.

- **Lock or unlock the control panel**: This may disable or enable the pole-mounted panel for crew members to conduct a regular service.

- **Battery test**: This is to evaluate the battery life cycle remotely, to determine if on-site maintenance is required.

- **Manual fault indicator (FI) reset**: In the event of hardware malfunction or errors, this remote reset may be necessary in case the applications are not able to make a conclusive inference of the potential fault location.

Analog measurements provide the updated status of a distribution feeder as to how much power is flowing through each FRTU node. This is linked with the SCADA addresses to all FRTUs in the control center that are useful for computation. Most of the pole-mounted FRTUs come with the basic measurement units, with a programmable logic controller (PLC) in the panel, described *in phases* as follows:

- Phase/neutral currents

- Phase voltages

- Daily peak load currents

- Daily statistics of peak and average of load currents

- Power factor

- Apparent power

- Other derived units, such as active and reactive power, energy, etc.

The following digital (binary) measurements are typically modeled in the SCADA data communication link to the FRTU which always includes additional quality code of communication unavailability:

- **Fault indicators**: This is a binary status to indicate whether or not the FRTU sees that there is a fault, with a simple binary indication that provides the additional information to help the fault localization module in the control center infer the potential faulted segment of a feeder.

- **Switch status**: The FRTU/FTU is wired to connect a swiching device, such as a breaker, recloser, or sectionalizer. The basic three quality codes for telemetered switches are: (1) open, (2) closed, and (3) in-transit. The connection of the switch with FRTU/FTU makes it an automated (remote-controlled) switch.

- **Battery and case door statuses**: These help operators inform the crew to work on battery replacement, or if a panel door is tampered with physically. This is an indicator to ensure the physical security of the panel.

- **Open line detection**: This determines the state of a line associated with the FRTU as to the energization status on both terminals.

- **Phase synchronization failure**: This logical check will determine if two or more cyclic signals tend to oscillate with relative phasor differences in a consistent waveform.

The second-by-second acquisition is measured and centralized to the distribution control center. With today's applications, the importance of capturing the latest states of system health is done through measurements and up-to-date GIS topology, to be frequently updated in the SCADA network. For the remainder of this chapter, the potential applications deployed in today's computerized systems for distribution networks are briefly discussed. These applications provide a representation of today's distribution management systems and may not cover specific niches. Applications can vary from vendor to vendor.

1.2.2 Extension with DMS Applications

DMS applications are the SCADA extensions for distribution automation that computerize and automate outage management functions, to coordinate available crew personnel in searching for possible paths from tie switches, i.e., NO switches, based on fault area using real-time measurements from the FRTUs/RTUs. Depending upon the availability of trouble call tickets, the search may or may not include call tickets associated with distribution transformers. An advanced version of the search would integrate with IP-based energy meters (often called smart meters) or other communication means to replace a missing notification from associated customers. Typically, the core functions of DMS include the network application with basic unbalanced power flow that receives information from topology processors as a base case, used to evaluate other "what-if" scenarios for possible fault isolation and power restoration of the "healthy" subsystem from other feeder(s) through NO switches. An integral part of the computerized applications would be twofold: (1) network analysis and (2) outage management. These two parts are highly dependent upon each other to provide necessary information that can help to dispatch a crew to the site as well as generating automatic switching procedures from topology processors and power-flow modules. The following applications are generally promoted in most DMS applications:

- **GIS importation**: The GIS importation module establishes a topological database in SCADA systems that are extracted from GIS datasets that are related to an electrical distribution network. This is the means to extract electrical information from geographical locations that will establish a graph for the remainder of the book. The topology will be used to navigate nodes and simplification of topology for the search of faulted areas.

- **Unbalanced radial power flow**: The input of this module depends upon load allocation from the telemetered measurements, as well as forecasts with updated topology status, to determine the states of each fictive node. This is the core module that will be related to other applications for power restoration. The recommendation by applications with the power flow losses and voltage estimates from the measurements of topology status is useful for operators in making a decision about what options would be optimal before the reconfiguration. The variables considered in the decision-making process are often for comparison purposes, to see which option would have optimal power losses or/and the least violation in voltage profiles.

- **Short circuit calculation**: This is an offline study of short-circuit calculation for a protection coordination study with other circuit breakers or reclosing devices as well as fuses. This is often used in planning where the operators could decide would be the optimal placement of new reclosers to capture the possible faults within a feeder. As feeders can be interconnected with normally open (NO) switches, the possibilities sites serve for the overlapping areas that would capture the overcurrents for protection coordination.

- **Fault localization, isolation, and service restoration**: Feeders are sectionalized with switches that can be remote controlled. This module is to temporarily rearrange power outage. The localization of a fault in order to isolate the segment will take place first. The temporary arrangement to restore power to "healthy" segments form other source(s) of the feeder.

- **Outage management and crew coordination**: This module relates to scheduled and unscheduled power outages that connect to other modules to narrow down the smallest faulted segment during the search with crew utilization in fault searches.

- **Trouble call and customer information system**: During a search for faults, trouble-call tickets would be useful to confirm outage customers as well as to investigate multiple locations of faults (independent events). As their customer identity and address will relate to a distribution transformer, the exploration of outage associated with prioritized customers would help in escalating the process of power restoration.

- **Generation of switching procedures**: Switching steps can be generated automatically or manually. The automatic generation of switching sequences would be based on the topological statuses. For safety reasons, the manually generated steps could also be heuristic such that knowledge would include some interlocking rules and design.

1.2.3 Coordinating Crew Location Using Mobile Apps

The most conventional way of locating a fault segment is the exhaustive search of a large area. Although there may be other techniques to trace and infer estimated fault location from the sending substation, the actual identification of faults can remain extremely time-consuming. Hence, providing real-time location of crew members would be useful information to coordinate search efforts and effectively identify fault location. The mobile apps can report crew position and other members to the distribution dispatching center. The geographical maps that overlap electrical topologies with a transportation layer would provide an intuitive assimilation to the operators in identifying the efforts in the search process. This is the extension to DMS applications that would expedite fault search time and would result in improving the overall system reliability. The search time can be tremendously reduced if the preliminary identification of a potential fault segment is identified from a fault indicator in the SCADA network. However, having more than one faulted circuit may occur from a single disturbance.

1.3 Customer Billing Center

Historically, utility crews will regularly visit customers' sites to read electromechanical analog energy meters for billing purposes. These datasets are manually updated in the relational database on a monthly basis. Such labor-intensive reading can be inefficient and is now gradually being replaced with technology-enabled communication, i.e., AMI, in which each household is deployed with a smart meter for the same purpose with frequency of up to one snapshot every ten minutes in kWh. The electronic device can also allow the distribution company to effectively infer potential outage areas or other abnormalities based on their regular reading. However, the drawback of deploying AMI can be cost-ineffective and is susceptible to cyberthreats.

The customer billing center is the corporate network that is separate from the distribution SCADA network. The customer billing databases may be set up with redundant servers for backup purposes. The customer ID could be the primary source of identification to relate to the distribution transformer ID and physical address of a household, which would be used to relate geographically and topologically in the SCADA /DMS network. The update of customer ID can be periodically exchanged under the script between the billing corporate network and SCADA network. Billing tools [38, 39] can be available to consumers but privacy has been a concern to energy users [40].

1.4   Backup Control Center and Data Replication

The communication of a distribution SCADA network is hierarchical where all pole-mounted device communications are centralized to the control center. Some utilities may implement backup control centers to ensure fail-over possibility to another site if there may be a disastrous event that may disable the communication infrastructure. Similarly, the data replication may be backed up locally within a site or in synchronization with another site. Utilities may also consider backup over the Internet using cloud storage with an enhanced encryption mechanism in order to assure privacy of customers is well protected during synchronization between two networks.

1.5   Conclusions

With the development of distribution SCADA networks and/or metering infrastructures, it is possible to accurately analyze and model distribution systems. This introductory chapter provides an overview of communication network centers that harness the telemetered information from primary and secondary distribution networks. The metering devices are mainly the pole-mounted devices, RTUs, and FRTUs, in the primary distribution network, and the smart meters on the customers' side, which is on the secondary distribution network. The customer billing center introduces demand response to adjust the peak and non-peak power usage according to the measurements feedback from metering devices, and sets the dynamic pricing rules. The two-way communication between the meter and the supplier in smart meters encourages customers to adjust their consumption habits to be more responsive to the dynamic electricity prices. Although the metering infrastructure coverage in the U.S. power utilities has risen, the rate of deploying metering devices remains at a slow pace, and modeling individual load consumption accurately within a distribution feeder has been challenging due to the limited metering points in the networks.

Advanced metering infrastructure (AMI) provides direct metering from the smart meter directly to the customer billing center where it is aggregated by a different network of the SCADA operational network. The customer billing center, which is part of the information technology (IT) network, has limited access to the SCADA network, but the billing-related information can be linked to the operational entities, such as customer billing information that connects directly to the distribution transformers (DTs) where the power comes from. This will be discussed further in Chapter 7, which discusses outage coordination and correlation to infer outage segments between switches. With a mobile app for the crew that is investigating the fault at a site, the exchange of real-time information between them, the control centers, and the customer billing center may expedite the search time, saving time as a result and tremendously improving the reliability of the system.

Discussion: Feeder Complexity

Computerized management systems appear to be part of today's automation and can improve the overall system reliability. The complexity of distribution networks can be simplified with fewer monitoring devices. Due to the geographically dispersed IP-based devices to be deployed, it is obvious that the initial deployment and ongoing maintenance costs can be a burden to utilities. Imagine a feeder with 800 distribution transformers and 50 sectionalizers. How would you decide to convert those sectionalizers into remote-controlled switches? At some point, inclusion of a recloser downstream may be necessary due to the long distance of distributing power to consumers. On what criteria should you decide? Based on individual feeders? Or depending upon the possible reconfiguration possibilities with other normally open (NO) tie switches with other feeder(s)? If there is a cost constraint, what decision variables make it a higher priority? Obviously, if it is otherwise, it would have been nice to have all 80 sectionalizers upgraded to be remote controllable. What other uncertainties might be involved in maintaining these IP-based communication methods and infrastructure?

Part II

Data Preparation and Modeling for Operation

2

Graph Modeling of Interconnected Feeders

The subject of grid topology has been the foundation of power flow for transmission networks where the Y matrix formation has been promoted in many textbooks that are embedded with adjacency and degree matrices. The Y matrix formation is also known as a Laplacian matrix with $(n \times n)$ size. In graph theory, the topology can be represented by either adjacency or incidence matrices. This chapter revisits the fundamental principles of graphs.

Due to the radial and unbalanced nature of distribution networks, the topology can be represented in an incidence matrix where the topological status and connectedness of interconnected feeders is crucial to the applications in the control center. The architecture of distribution SCADA/DMS is built based on the customer interaction management modeling where the topology is salient to the construct for application usage. Kersting's notion on "garbage in, garbage out" essentially refers to outdated topology status where the computational results would not reflect the real-world scenarios [41].

In this chapter, we establish the fundamentals of graph representation in a distribution network where all feeders can be interconnected between/within substations with a tie switch. The interdependencies of substation connectivity and possibilities for reconfiguration are essential to the structure of node navigation of a graph when there is a fault within a network.

Distribution feeders are a graph that is mathematically structured with pairwise relationships between objects. The context of a graph in this book is the relationship between electrically connected objects, which can be represented by two basic components: a finite set of vertices (also called nodes, fictive buses). A set of the pairs of nodes forms an edge. At least two vertices will form an edge. An edge is the component that is used to represent the one-way or two-way relationship between two vertexes in a graph. The edges may contain weight, value, and/or cost [42–45].

The topology of a graph is often used to establish the flow of information [46,47]. In the context of this book, we will use it to represent power flow direction, energization states of a feeder, and the connectedness of a feeder as well as its interdependencies with other feeders. The sources modeled in this book are the coupling points of power injection from the sub-transmission system through a distribution substation. However, one may argue that a distributed energy resource can be used to model a negative load that represents the generating sources of power within a feeder. The main focus of this book is to establish the aforementioned structures of topological representation.

2.1   Graph Representation

Generally speaking, a distribution feeder consists of a finite set of vertices
and edges that may have line segments, disconnectors, Capacitors, loads, and
voltage regulators that are energized from a feeder head connected to a distribution substation. This chapter establishes the nomenclature of notational
consistencies as well as the expected structure of subsystem topological statuses for the remainder of the other chapters. We divide this section into three
parts:

1. Representation of distribution elements,

2. Formats of matrices, and

3. Traverse algorithms.

The graph representation is where we establish the models from distribution elements. The formats of matrices include adjacency and incidence matrices [42] and how we will organize the vertices and edges of those elements.
The traverse algorithms provide the basic reorganization of matrix formats,
which include numbering as well as the possible pre-arranged flow direction
of power. These three topics include the critical elements and formats to be
elaborated in the later chapters, where they are used to determine subsystems
and energization states.

2.1.1   Representation of Distribution Elements

A distribution feeder typically consists of hundreds of electrical elements in
the network. The following enumerates the essentials of typical distribution
elements in a feeder [41].

1. Line segments (edges)

2. Distribution transformers (edges)

3. Switches (edges)

4. Voltage regulators (edges)

5. Loads (vertices)

6. Capacitors (vertices)

7. Distributed energy resources (vertices)

Most of them are represented in edge type in a graph. Electrical loads, capacitors, and distributed energy resources are represented in *vertices*. Lines,
distribution transformers, switches, and voltage regulators are modeled as

edges in a graph. A set of these elements forms a feeder that is usually energized in a radial form, i.e., it is generally referred to as a spanning tree in graph theory [48]. However, the representation of a feeder with other feeders and the possibilities are in mesh form and are partitioned with NO switches. The mesh form is illustrated as a multiple loop in an undirected manner.

2.1.2 Formats of Adjacency and Incidence Matrices

Table 2.1 provides a comparison between adjacency and incidence matrices. The matrix formats are the adjacency and incidence matrices. The two forms of matrices can directly and indirectly provide information on the fly. Incidence matrices are the ones that will be fully used for this book because they provide edges and vertices information to show the network connectivity that can be visualized and related in an intuitive manner. However, it does not mean that the adjacency matrices do not provide edge information. The edge information can be derived by itself. This is referred to as indirect information that requires steps to convert. Either format can represent direction information. The adjacency matrices are always a square matrix but their symmetry depends on how they are modeled. An undirected adjacency matrix is symmetric but a directed adjacency matrix is not. Incidence matrices are not square unless the number of edges and vertices are the same. They are never symmetric [49, 50].

In terms of memory storage, both adjacency and incidence matrices can be highly sparse graphs as the size of the matrix grows larger with the zeros [51]. The connectivity is represented by 1's in the network. The radial network may only appear in a few connected elements in each node where the remainder of the columns are not connected to any other node. The non-matrix elements, i.e., 1's are stored in memory. In MATLAB, the optimal way to store the non-zero elements can be archived using the `sparse` function [50, 52]. In general, the adjacency matrix is relatively sparser than the incidence matrix.

Table 2.1 enumerates only a directed incidence matrix consisting of -1 that provides the flow direction of power. The adjacency matrix represents a directed graph as an asymmetrical graph. On the other hand, the undirected graph on both representations can be with the elements of 0 or 1. Such a format represents the connectedness between vertices.

TABLE 2.1

Comparison Between Adjacency and Incidence Matrices

	Adjacency Matrix		Incidence Matrix	
Types	Undirected	Directed	Undirected	Directed
Square Matrix	Yes	Yes	No	No
Sparsity	Higher	Higher	Mostly Less	Mostly Less
Element Types	0, 1	0, 1	0, 1	0, 1, -1

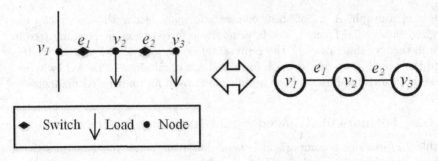

FIGURE 2.1
A simple feeder example with its equivalent graph representation.

Illustrated in Fig. 2.1 is a simple 3-node feeder with 2 edges where those edges are switches. This feeder is connected with 2 loads. In graph representation, this is the scenario that consists of 3 vertices and 2 edges. The formats of matrices for the undirected and directed graphs are shown below. The following directed and undirected adjacency matrices are the equivalent graph for the simple feeder example:

$$
\begin{array}{c}
\begin{array}{ccc} v_1 & v_2 & v_3 \end{array} \\
\begin{array}{c} v_1 \\ v_2 \\ v_3 \end{array}
\begin{pmatrix}
0 & 1 & 0 \\
0 & 0 & 1 \\
0 & 0 & 0
\end{pmatrix}
\end{array}
\qquad
\begin{array}{c}
\begin{array}{ccc} v_1 & v_2 & v_3 \end{array} \\
\begin{array}{c} v_1 \\ v_2 \\ v_3 \end{array}
\begin{pmatrix}
0 & 1 & 0 \\
1 & 0 & 1 \\
0 & 1 & 0
\end{pmatrix}.
\end{array}
$$

Similarly, the directed and undirected incidence matrices are shown below:

$$
\begin{array}{c}
\begin{array}{cc} e_1 & e_2 \end{array} \\
\begin{array}{c} v_1 \\ v_2 \\ v_3 \end{array}
\begin{pmatrix}
-1 & 0 \\
1 & -1 \\
0 & 1
\end{pmatrix}
\end{array}
\qquad
\begin{array}{c}
\begin{array}{cc} e_1 & e_2 \end{array} \\
\begin{array}{c} v_1 \\ v_2 \\ v_3 \end{array}
\begin{pmatrix}
1 & 0 \\
1 & 1 \\
0 & 1
\end{pmatrix}.
\end{array}
$$

By observation of these two forms of graph representation, a connection between two nodes is established when the interception of row and column has a 1 in it, e.g., v_1 and v_2 are connected with a 1 in adjacency matrix. However, such connection between the two nodes is represented in column-overlapping, i.e., v_1 and v_2 both have a 1 in column e_1. By comparing the directed adjacency and incidence matrices, the adjacency matrix only shows 1 as an indication of entering a node while the incidence matrix has 1 and -1, which illustrates entering and leaving the node as directional flow.

These two forms of matrices can be converted from one or the other directly. There exist multiple conversion methods in the graph theory literature based on (1) concept, (2) structure, and (3) connectivity [43–45, 50, 53]. As such conversion may be used frequently throughout the book, the pseudocode

---

Algorithm 1 Structure-Based Adjacency and Incidence Matrix Conversion Algorithm

---

Require:
  Set M as input adjacency or incidence matrix.
  Set c as a reference coefficient,
    where 0 represents a conversion from adjacency matrix to incidence matrix;
        1 represents a conversion from incidence matrix to adjacency matrix.
  Set R as the result stack.
Ensure: MatrixConversion(M, c)
  if $c = 0$ **then**
    Calculate the number of edges m;
    Calculate the number of nodes n;
    Initial $k = 1$ as the edge index and $k \in [1, m]$;
    for $i \leftarrow 1 : n$ **do**
      for $j \leftarrow i : n$ **do**
        if $M(i, j) \neq 0$ **then**
          $R(i, k) \leftarrow 1$
          $R(j, k) \leftarrow 1$
          $k \leftarrow k + 1$
        end if
      end for
    end for
  end if
  if $c = 1$ **then**
    Calculate the number of edges m;
    Calculate the number of nodes n;
    for $i \leftarrow 1 : m$ **do**
      Find the index of nonzero element a column by column in M;
      $R(a(1), a(2)) \leftarrow 1$
      $R(a(2), a(1)) \leftarrow 1$
    end for
  end if
  return R

---

illustrates that the conversion is given in algorithm 1 and the corresponding MATLAB sample code is shown in Listing 2.1. This algorithm converts an undirected adjacency matrix into an undirected incidence matrix using the structure method.

The algorithm handles two ways of conversion: to convert from adjacency matrix to incidence or vice versa. The M is the input matrix; whether it is an adjacency or incidence matrix is to be determined. The c defines the matrix conversion type. As this algorithm is two-part, defining c will lead to different handling of conversion. The conversion involves reorganizing the orientation matrix that is described in the algorithm.

Listing 2.1
Structure-Based Adjacency and Incidence Matrix Conversion

```
1   % Structure based function for convertion between undirected ...
        adjacency matrix
2   % and undirected incidence matrix
3   % INPUT: F is the input matrix
4   % INDEX f = 0: adj2inc
5   % INDEX f = 1: inc2adj
6   % adj2inc example: Graph:    __(v1)<--
7   %                    /        \_e2/e4_
8   %               e1|                   |
9   %                 \-> (v2)-e3->(v3)<-/
10  %
11  %         AdjM = [0 1 1
12  %                 0 0 1
13  %                 1 0 0];
14  %
15  %             v1  v2 v3  <- vertices
16  %             |   |  |
17  %         IncM = [1 -1  0   <- e1   |
18  %                 1  0 -1   <- e2   | edges
19  %                 0  1 -1   <- e3   |
20  %                -1  0  1]; <- e4   |
21  %================================================================%
22  % inc2adj example: Graph:    __(v1)<--
23  %                    /        \_e2/e4_
24  %               e1|                   |
25  %                 \-> (v2)-e3->(v3)<-/
26  %
27  %             v1  v2 v3  <- vertices
28  %             |   |  |
29  %         IncM = [1 -1  0   <- e1   |
30  %                 1  0 -1   <- e2   | edges
31  %                 0  1 -1   <- e3   |
32  %                -1  0  1]; <- e4   |
33  %
34  %         AdjM = [0 1 1
35  %                 0 0 1
36  %                 1 0 0];
37  %================================================================%
38
39  function w = matrixconvert(F,f)
40
41  if f == 0
42      m = sum(sum(F))/2; % number of edges
43      n = size(F,1); % number of vertexs
44      w = zeros(n,m);
45      k = 1; %edges index k=1:m
46  for i = 1:n
47      for j =i:n %avoid checking the repeated nodes
48          if F(i,j) ≠ 0
```

```
49            w(i,k) = 1;
50            w(j,k) = 1;
51            k = k + 1;
52         end
53      end
54   end
55
56   elseif f == 1
57        m = size(F,2);
58        n = size(F,1);
59        w = zeros(n,n);
60      for i = 1:m
61          a = find(F(:,i) ≠ 0);
62          w(a(1),a(2)) = 1;
63          w(a(2),a(1)) = 1;
64      end
65   else
66        fprint('f valut is NOT correct!')
67   end
68   w;
69   end
```

It is important to note that the topology of graph representation in the circuit is equivalent to the admittance Y-bus matrix, which is also called the Laplacian matrix [54, 55]. The Laplacian matrix is frequently applied in electrical circuits and analysis, particularly in the Kron reduction process, and various applications such as electrical impedance tomography, the sensitivity of reduced power flow, and transient stability assessment in power networks [56]. The Laplacian matrix can only be applied to an undirected graph without loops and can be obtained from Eq. 2.1, where L is the Laplacian matrix, D is the degree matrix (diagonal elements), and A is the adjacency matrix (off-diagonal elements).

$$L = D - A. \qquad (2.1)$$

Shown in Fig. 2.2 is an example of an undirected graph with 6 vertices. The corresponding degree matrix and the adjacency matrix are also demonstrated as below. The Laplacian matrix is a symmetrix matrix. The off-diagonal entries are nonpositive and they are either 1 or -1, while the diagonal elements are positive. The diagonal entries are the vertex degrees and the row sums and the column sums are all zero. A Laplacian matrix (embedded with an adjacency matrix) is frequently used in transmission power flow analysis because it can be established by inspection. As the matrix is of equal dimensions and each element of the Y admittance matrix represents the direct connectivity between nodes, we simply take full advantage of such features to form the mesh network that is updated by accessing the symmetric connection information, element by element, from the non-sparse values based on the orientation. Such manipulation is relatively straightforward in handling and debugging topologically for mesh transmission networks.

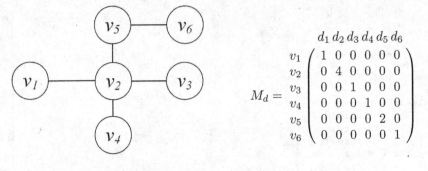

$$M_d = \begin{array}{c} \\ v_1 \\ v_2 \\ v_3 \\ v_4 \\ v_5 \\ v_6 \end{array} \begin{array}{cccccc} d_1\ d_2\ d_3\ d_4\ d_5\ d_6 \\ \left(\begin{array}{cccccc} 1 & 0 & 0 & 0 & 0 & 0 \\ 0 & 4 & 0 & 0 & 0 & 0 \\ 0 & 0 & 1 & 0 & 0 & 0 \\ 0 & 0 & 0 & 1 & 0 & 0 \\ 0 & 0 & 0 & 0 & 2 & 0 \\ 0 & 0 & 0 & 0 & 0 & 1 \end{array}\right) \end{array}$$

$$M_a = \begin{array}{c} \\ v_1 \\ v_2 \\ v_3 \\ v_4 \\ v_5 \\ v_6 \end{array} \begin{array}{cccccc} v_1\ v_2\ v_3\ v_4\ v_5\ v_6 \\ \left(\begin{array}{cccccc} 0 & 1 & 0 & 0 & 0 & 0 \\ 1 & 0 & 1 & 1 & 1 & 0 \\ 0 & 1 & 0 & 0 & 0 & 0 \\ 0 & 1 & 0 & 0 & 0 & 0 \\ 0 & 1 & 0 & 0 & 0 & 1 \\ 0 & 0 & 0 & 0 & 1 & 0 \end{array}\right) \end{array} \quad M_l = \left(\begin{array}{cccccc} 1 & -1 & 0 & 0 & 0 & 0 \\ -1 & 4 & -1 & -1 & -1 & 0 \\ 0 & -1 & 1 & 0 & 0 & 0 \\ 0 & -1 & 0 & 1 & 0 & 0 \\ 0 & -1 & 0 & 0 & 2 & -1 \\ 0 & 0 & 0 & 0 & -1 & 1 \end{array}\right)$$

FIGURE 2.2
Laplacian matrix example.

2.1.3 Traverse Algorithms

Traversal of a graph is important to navigate the nodes of a given topology. An update of a subsystem based on switching execution within related feeders would require an update of the directional flow of power, which will be discussed in more detail in the chapters on distribution management applications. Graph traversal updates every visited vertex and edge with an ordering number, and its efficiency would depend on the algorithms to handle the structure of a topology. Generally, the start of nodal traversal begins at the feeder head from a substation to the rest of the feeder, which would end at tie switches. We refer to tie switches as NO switches throughout this book. As the direction of power flow would be determined by the power sources as well as possibly distributed energy resources, the traversal may include the direction of preliminary update based on directed graph theory. A general representation of most common feeders of an electrical network, in terms of traversal or searching within a subnetwork, is as follows:

1. Determine whether a subsystem is a directed acyclic graph (DAG) [43–45];

2. Sort topological ordering in a network numbering for subsystems;

3. Identify the best location of the root node for a radial topology as the distribution substation;

4. Ensure all associated nodes and loads are reachable; and

5. Identify all possible shortest paths within a feeder for optimal system reconfiguration during a faulted event.

All of these applications are based on real-time SCADA information to update the topology based on injection measurements and switching statuses, as well as other indicators such as a fault or alarm indicator.

Algorithm 2 Breadth-First Search Algorithm

Require:
    Set M_a as the input adjacency matrix;
    Set *queue* as the search queue;
    Initial the queue head *head* = 1;
    Initial the queue tail *tail* = 1,
    then the queue is empty because *head* = *tail*;
    Set *flag* as the mark to check the discovered status;
    Set R as the result stack;
Ensure: BFS(A)
    Calculate how many vertices m are in this topology;
    Start searching the queue from the first vertex (root node);
    Search the next element in the queue;
    while *queue* is not empty **do**
      $i \leftarrow$ the last vertex in the queue;
      for $j \leftarrow 1 : m$ **do**
        if vertices are connected and undiscovered **then**
          queue(*head*) $\leftarrow j$;
          head $\leftarrow$ *head* + 1;
          flag $\leftarrow j$;
          $R \leftarrow [i, j]$;
        end if
      end for
      tail $\leftarrow$ *tail* + 1
    end while
    return R

2.1.3.1 Breadth-First Search

Breadth-first search (BFS) is commonly used in graph traverse algorithms [42–45] to navigate nodes within a subsystem of a feeder. The traversal starts from a selected node (typically, it is a node of the power injection source in the distribution substation) to the children nodes associated with the root node. This exploration continues to the next depth levels of nodes until the end of the spanning tree. An immediate application in distribution engineering is backward-forward distribution power flow analysis where each level of backward or forward would depend on the sibling nodes' results. This algorithm is utilized to ensure all load nodes are reachable under an energized subsystem.

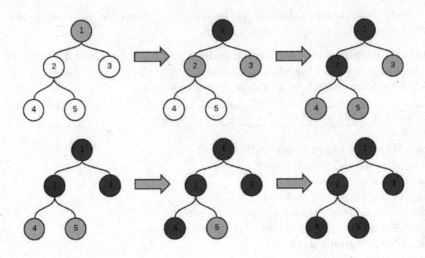

FIGURE 2.3
Search process of the breadth-first search algorithm.

Algorithm 2 shows the breadth-first search applied in an adjacency matrix. This algorithm can be demonstrated with an example 5-node graph shown in Fig. 2.3. The navigation of the graph starts from the root node, which is node 1 in the graph. The white nodes are yet to be traversed, gray nodes are the same level of siblings, and the dark nodes are discovered (counted). The sequential queue of this example then becomes 1, 2, 3, 4, 5. This order is often treated as an updated topology.

2.1.3.2 Depth-First Search

Like the BFS method, the depth-first search (DFS) is typically used for topological sorting in a given graph [42–45, 57]. For the applications of distribution management systems, the obvious use of this method is to search for the "shortest path." This is referred to as the system reconfiguration during fault occurrence where the algorithm will search for possible paths of power sources. The sources are usually from the injection points from other feeders within the same substation or from other substations connected with normally open switches. We will introduce such applications in the later chapters on crew management and coordination as well as temporary power restoration to resume power for the "healthy" segment of a faulted feeder. The optimization of the possibilities would rely on the total power losses (in terms of distance) and interdependent sources from other substations as well as information about possible overloads. Fig. 2.4 shows how DFS navigation takes place. This is similar to BFS, but the priority of exploring nodes goes by depth level first. The queue of this example is 1, 2, 4, 5, 3.

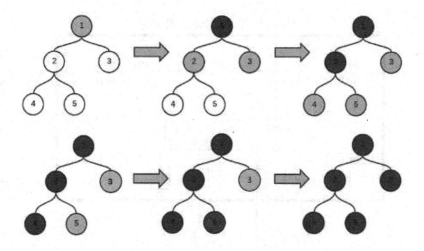

FIGURE 2.4
Search process of the depth-first search algorithm.

2.1.3.3 Topological Ordering

The topological sorting or (re-)ordering of directed graphs is a row vector of visited nodes in a sequential order [42–45]. For each directed edge e_{ij} from the vertex i to the vertex j, i is preceded by j in each element of the row vector.

In task assignment, an ordered graph could represent the tasks to be performed, where each edge can represent a constraint, and the ordered numbers indicate the priority of the tasks, which are sorted in a sequential manner. Another example would be applied in an electrical distribution system. The sorting would start from the feeder head of an injection point to the rest of the subsystems. The reordering would ensure optimality of navigation, which also determines the topological state of a given subsystem, e.g., electrically energized in a loop, mesh, or radial structure. It also can provide an update on directional flow of power. All visited nodes in a graph are not necessarily assigned with a sign of direction. However, if a from-to direction is assigned from the power flow result, it is a DAG. The DAG then has sorted at least once topologically if (i) each vertex only appears once and (ii) there is no path from vertex B to vertex A if the order of vertex A is ahead of vertex B. A new update of DAG would imply an incremental update of a given graph. Kahn's algorithm [58] is one way of nodal exploration to determine only all leaving vertices of a graph.

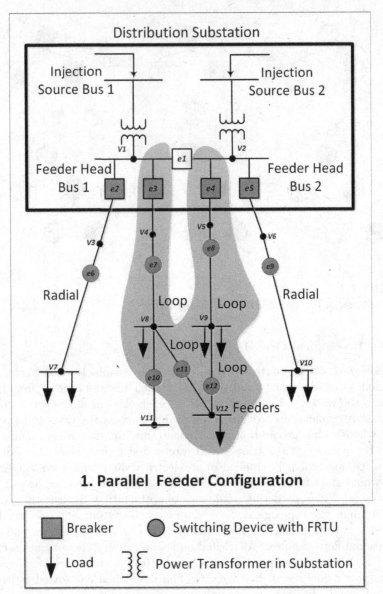

FIGURE 2.5
Parallel feeder configuration.

2.1.3.4   Parents and Children Function

In the graph representation of an electrical distribution network, the power
injection source and the connected nodes and loads of a node can be found by

searching parents and children nodes of the target node in the given topology. A root node represents feeder head, and multiple feeder heads are associated with a substation. The input of this function is the adjacency matrix. The parent node can be found by searching the index of the first non-zero element in the column of the object node in the given adjacency matrix, while the index of the second non-zero element represents the child node.

An example is shown in Fig. 2.6 and the corresponding adjacency matrix is following the topology. Assuming node 5 as the target node, the first non-zero element in the fifth column appears in the first row and the second non-zero element in the sixth row. According to the figure, the parent node of node 5 is node 1 while the child node of node 5 is node 6.

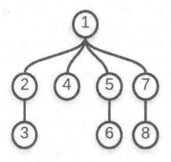

$$
\begin{array}{c@{\quad}c}
 & \begin{array}{cccccccc} 1 & 2 & 3 & 4 & 5 & 6 & 7 & 8 \end{array} \\
\begin{array}{c} 1 \\ 2 \\ 3 \\ 4 \\ 5 \\ 6 \\ 7 \\ 8 \end{array} &
\left(\begin{array}{cccccccc}
0 & 1 & 0 & 1 & 1 & 0 & 1 & 0 \\
1 & 0 & 1 & 0 & 0 & 0 & 0 & 0 \\
0 & 1 & 0 & 0 & 0 & 0 & 0 & 0 \\
1 & 0 & 0 & 0 & 0 & 0 & 0 & 0 \\
1 & 0 & 0 & 0 & 0 & 1 & 0 & 0 \\
0 & 0 & 0 & 0 & 1 & 0 & 0 & 0 \\
1 & 0 & 0 & 0 & 0 & 0 & 0 & 1 \\
0 & 0 & 0 & 0 & 0 & 0 & 1 & 0
\end{array}\right)
\end{array}
$$

FIGURE 2.6
Parents and children function example.

2.2 Examples for Topologies of Distribution Primary Network

Each distribution substation is branched out with multiple feeders that are typically energized in a radial structure. Under the same system of a graph representation used in the previous sections, we enumerate all possible unique topological configurations that may be used later for illustrating an operational scenario. Some cases may be transitional states, such as to isolate a fault. A feeder should first be isolated before connecting a feeder from other substations or the same substation. The energization state of the topological status would provide a transition to power restoration, i.e., to those "healthy" parts of the segments restored from other power sources. The complete enumeration of unique cases is:

1. Parallel feeder configuration (under the same substation)

2. Loop with other feeder configuration (under the same substation)

3. Loop within a feeder configuration (weakly meshed)

4. Parallel and loop feeder configuration with the same substation as well as another substation

5. Parallel feeder configuration (with the other substation)

2.2.1 Parallel Feeder Configuration

As shown in Fig. 2.5, the configuration is possible when a substation has more than one feeder. It is usually a temporary arrangement to transition from one state to the other. In the graph representation, one can visualize the incidence matrix into 3 colored clusters (see color e-book). The reliability of this configuration is supported by the two injection sources from the substation, which are stepped down by two independent power transformers. The tie breaker between the two transformers, denoted as e_1, relates these two subsystems so that if one of the power transformers fails, the other would pick up the loads through the tie breaker.

The following incidence matrix describes the topology of the given figure. Each column represents the number of vertices that consistently show up in the figure. Similarly, each row represents the edges of each element segment. The M_i of a parallel feeder system is expressed as:

$$
M_i = \begin{array}{c} \\ v_1 \\ v_2 \\ v_3 \\ v_4 \\ v_5 \\ v_6 \\ v_7 \\ v_8 \\ v_9 \\ v_{10} \\ v_{11} \\ v_{12} \end{array}
\begin{array}{c} e_1\ e_2\ e_3\ e_4\ e_5\ e_6\ e_7\ e_8\ e_9\ e_{10}\ e_{11}\ e_{12} \\
\left(\begin{array}{cccccccccccc}
1 & 1 & 1 & 0 & 0 & 0 & 0 & 0 & 0 & 0 & 0 & 0 \\
1 & 0 & 0 & 1 & 1 & 0 & 0 & 0 & 0 & 0 & 0 & 0 \\
0 & 1 & 0 & 0 & 0 & 1 & 0 & 0 & 0 & 0 & 0 & 0 \\
0 & 0 & 1 & 0 & 0 & 0 & 1 & 0 & 0 & 0 & 0 & 0 \\
0 & 0 & 0 & 1 & 0 & 0 & 0 & 1 & 0 & 0 & 0 & 0 \\
0 & 0 & 0 & 0 & 1 & 0 & 0 & 0 & 1 & 0 & 0 & 0 \\
0 & 0 & 0 & 0 & 0 & 1 & 0 & 0 & 0 & 0 & 0 & 0 \\
0 & 0 & 0 & 0 & 0 & 0 & 1 & 0 & 0 & 1 & 1 & 0 \\
0 & 0 & 0 & 0 & 0 & 0 & 0 & 1 & 0 & 0 & 0 & 1 \\
0 & 0 & 0 & 0 & 0 & 0 & 0 & 0 & 1 & 0 & 0 & 0 \\
0 & 0 & 0 & 0 & 0 & 0 & 0 & 0 & 0 & 1 & 0 & 0 \\
0 & 0 & 0 & 0 & 0 & 0 & 0 & 0 & 0 & 0 & 1 & 1
\end{array}\right)
\end{array}.
$$

2.2.2 Loop with Other Feeders Configuration

Generally, the distribution network is operated in the radial or open loop due to the limitations of safety and protection. Since the power flow on feeders is in one direction only, the radial operation allows the use of a protection system without directional discrimination. However, with the high-quality demand for power supply and the improved service requirements, as well as the rapid expansion of modern power systems, the closed loop operation has been adopted gradually. Except for decreasing power losses during system control and management, closed loop operation can also keep feeder voltage level with a higher capacity for load rising.

As shown in Fig. 2.7, the configuration has a loop within two feeders and the closed loop is visualized as 2 colored clusters (see color e-book) in the corresponding incidence matrix. Compared with open loop, the tie switch e_1 is closed naturally in this operation so that a fault isolation on one section (node or line segment) within the loop cannot cause the power outage. The closed loop operation is transient and under the distribution network constraints.

The incidence matrix in the loop connecting the other feeders is:

$$
M_i = \begin{array}{c}
\\ v_1 \\ v_2 \\ v_3 \\ v_4 \\ v_5 \\ v_6 \\ v_7 \\ v_8 \\ v_9 \\ v_{10} \\ v_{11} \\ v_{12}
\end{array}
\begin{pmatrix}
\begin{array}{cc|cccccc|cccc}
e_1 & e_2 & e_3 & e_4 & e_5 & e_6 & e_7 & e_8 & e_9 & e_{10} & e_{11} & e_{12} \\
1 & 1 & 1 & 0 & 0 & 0 & 0 & 0 & 0 & 0 & 0 & 0 \\
1 & 0 & 0 & 1 & 1 & 0 & 0 & 0 & 0 & 0 & 0 & 0 \\
0 & 1 & 0 & 0 & 0 & 1 & 0 & 0 & 0 & 0 & 0 & 0 \\
0 & 0 & 1 & 0 & 0 & 0 & 1 & 0 & 0 & 0 & 0 & 0 \\
0 & 0 & 0 & 1 & 0 & 0 & 0 & 1 & 0 & 0 & 0 & 0 \\
0 & 0 & 0 & 0 & 1 & 0 & 0 & 0 & 1 & 0 & 0 & 0 \\
0 & 0 & 0 & 0 & 0 & 1 & 0 & 0 & 0 & 0 & 0 & 0 \\
0 & 0 & 0 & 0 & 0 & 0 & 1 & 0 & 0 & 1 & 1 & 0 \\
0 & 0 & 0 & 0 & 0 & 0 & 0 & 1 & 0 & 0 & 0 & 1 \\
0 & 0 & 0 & 0 & 0 & 0 & 0 & 0 & 1 & 0 & 0 & 0 \\
0 & 0 & 0 & 0 & 0 & 0 & 0 & 0 & 0 & 1 & 0 & 0 \\
0 & 0 & 0 & 0 & 0 & 0 & 0 & 0 & 0 & 0 & 1 & 1
\end{array}
\end{pmatrix}
$$

2.2.3 Loop within a Feeder Configuration

This configuration may appear in a single feeder when there is a temporary operation by the crew for system maintenance or detection of the feeder segment. The advantage of this arrangement is that a power outage in one feeder cable will not lead to a part of the facilities being out of service. In addition, one feeder cable can be maintained or replaced without causing a loss of service [59]. As shown in Fig. 2.8, the loop is shown as a colored cluster (see color e-book) in the incidence matrix. The reliability and security of this configuration are supported by the power injection from the feeder head. The

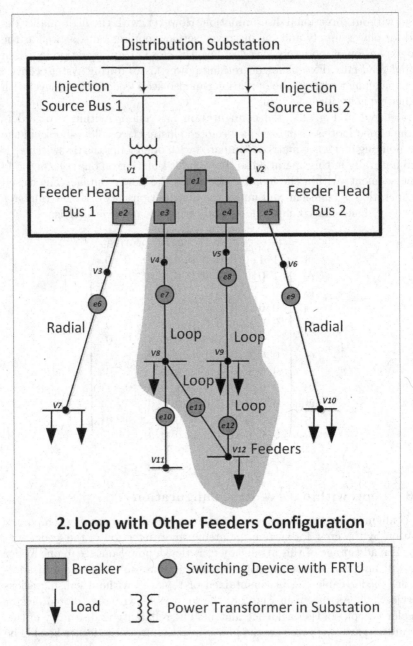

FIGURE 2.7
Loop with other feeder configuration.

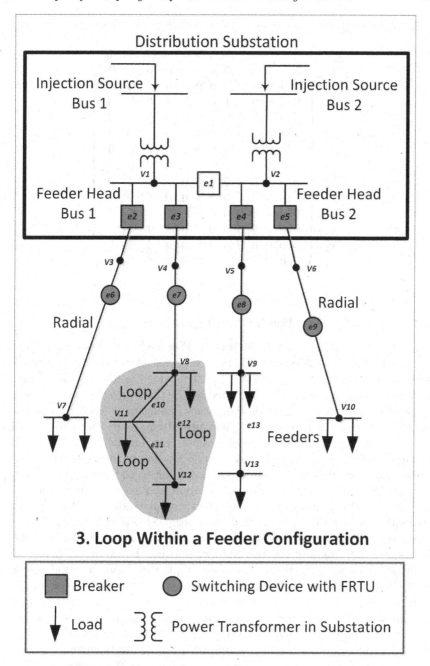

FIGURE 2.8
Loop within a feeder configuration.

corresponding incidence matrix in the loop within a feeder configuration is:

$$
M_i = \begin{array}{c} \\ v_1 \\ v_2 \\ v_3 \\ v_4 \\ v_5 \\ v_6 \\ v_7 \\ v_8 \\ v_9 \\ v_{10} \\ v_{11} \\ v_{12} \\ v_{13} \end{array}
\begin{pmatrix}
\begin{smallmatrix} e_1 & e_2 & e_3 & e_4 & e_5 & e_6 & e_7 & e_8 & e_9 & e_{10} & e_{11} & e_{12} & e_{13} \end{smallmatrix} \\
1 & 1 & 1 & 0 & 0 & 0 & 0 & 0 & 0 & 0 & 0 & 0 & 0 \\
1 & 0 & 0 & 1 & 1 & 0 & 0 & 0 & 0 & 0 & 0 & 0 & 0 \\
0 & 1 & 0 & 0 & 0 & 1 & 0 & 0 & 0 & 0 & 0 & 0 & 0 \\
0 & 0 & 1 & 0 & 0 & 0 & 1 & 0 & 0 & 0 & 0 & 0 & 0 \\
0 & 0 & 0 & 1 & 0 & 0 & 0 & 1 & 0 & 0 & 0 & 0 & 0 \\
0 & 0 & 0 & 0 & 1 & 0 & 0 & 0 & 1 & 0 & 0 & 0 & 0 \\
0 & 0 & 0 & 0 & 0 & 1 & 0 & 0 & 0 & 0 & 0 & 0 & 0 \\
0 & 0 & 0 & 0 & 0 & 0 & 1 & 0 & 0 & 1 & 0 & 1 & 0 \\
0 & 0 & 0 & 0 & 0 & 0 & 0 & 1 & 0 & 0 & 0 & 0 & 1 \\
0 & 0 & 0 & 0 & 0 & 0 & 0 & 0 & 1 & 0 & 0 & 0 & 0 \\
0 & 0 & 0 & 0 & 0 & 0 & 0 & 0 & 0 & 1 & 1 & 0 & 0 \\
0 & 0 & 0 & 0 & 0 & 0 & 0 & 0 & 0 & 0 & 1 & 1 & 0 \\
0 & 0 & 0 & 0 & 0 & 0 & 0 & 0 & 0 & 0 & 0 & 0 & 1
\end{pmatrix}.
$$

2.2.4   Parallel/Loop Feeder Configuration

Fig. 2.9 shows an extension of the loop with other feeders configuration (Configuration 2). Another interconnection between two feeders fed by two different substations (Distribution Substations 1 and 2) is added as edge e_{14}. The power injection of this configuration is supported by the three injection source buses with three transformers. Also, this temporary arrangement is configured because the system may rely on two different substations; its performance should conform to system conditions and constraints. The highlighted area in this figure is visualized as 5 clusters in the corresponding incidence matrix. The positions and the number of clusters are according to the topological ordering result. Without tie switches, reconfiguration may result in outage as the healthy part of the segment, which is downstream, may not be fed with power from another source that is connected via a normally open (NO) switch.

This is the first scenario where there are interconnected feeders with tie switches and determination of the NO switches can be evaluated based on the optimality of network status. The states of parallel, loop, or radial can be formed when some tie switches are changed to NO, although one end of the switch must be energized in order to ensure the complete energization state fed by immediate feeder from a substation, and some parts are from another feeder. More than one normally closed (NC) switch from an injection source will result in a loop. Parallel status occurs when one feeder is electrically energized at the same time from two feeders that are not from the same substation. This is a topological status that is in transition to normal configuration and may only require opening the switches of e_{14} and e_{15} in order for all feeders to be radial.

The parallel/loop feeder configuration incidence matrix is:

$$
M_i = \begin{array}{c|ccccccccccccccccc}
 & e_1 & e_2 & e_3 & e_4 & e_5 & e_6 & e_7 & e_8 & e_9 & e_{10} & e_{11} & e_{12} & e_{13} & e_{14} & e_{15} & e_{16} & e_{17} \\
v_1 & 1 & 1 & 0 & 0 & 0 & 0 & 0 & 0 & 0 & 0 & 0 & 0 & 0 & 0 & 0 & 0 & 0 \\
v_2 & 1 & 0 & 0 & 1 & 0 & 0 & 0 & 0 & 0 & 0 & 0 & 0 & 0 & 0 & 0 & 0 & 0 \\
v_3 & 0 & 1 & 1 & 0 & 0 & 0 & 0 & 0 & 0 & 0 & 0 & 0 & 0 & 0 & 0 & 0 & 0 \\
v_4 & 0 & 0 & 0 & 1 & 0 & 0 & 0 & 0 & 0 & 0 & 0 & 0 & 0 & 0 & 0 & 0 & 0 \\
v_5 & 0 & 0 & 1 & 0 & 0 & 0 & 0 & 0 & 0 & 0 & 0 & 0 & 0 & 1 & 0 & 0 & 0 \\
v_6 & 0 & 0 & 0 & 0 & 1 & 1 & 1 & 0 & 0 & 0 & 0 & 0 & 0 & 0 & 0 & 0 & 0 \\
v_7 & 0 & 0 & 0 & 0 & 1 & 0 & 0 & 1 & 1 & 0 & 0 & 0 & 0 & 0 & 0 & 0 & 0 \\
v_8 & 0 & 0 & 0 & 0 & 0 & 1 & 0 & 0 & 0 & 1 & 0 & 0 & 0 & 0 & 0 & 0 & 0 \\
v_9 & 0 & 0 & 0 & 0 & 0 & 0 & 1 & 0 & 0 & 0 & 1 & 0 & 0 & 0 & 0 & 0 & 0 \\
v_{10} & 0 & 0 & 0 & 0 & 0 & 0 & 0 & 1 & 0 & 0 & 0 & 1 & 0 & 0 & 0 & 0 & 0 \\
v_{11} & 0 & 0 & 0 & 0 & 0 & 0 & 0 & 0 & 1 & 0 & 0 & 0 & 1 & 0 & 0 & 0 & 0 \\
v_{12} & 0 & 0 & 0 & 0 & 0 & 0 & 0 & 0 & 0 & 1 & 0 & 0 & 0 & 0 & 0 & 0 & 0 \\
v_{13} & 0 & 0 & 0 & 0 & 0 & 0 & 0 & 0 & 0 & 0 & 1 & 0 & 0 & 1 & 1 & 1 & 0 \\
v_{14} & 0 & 0 & 0 & 0 & 0 & 0 & 0 & 0 & 0 & 0 & 0 & 1 & 0 & 0 & 0 & 0 & 1 \\
v_{15} & 0 & 0 & 0 & 0 & 0 & 0 & 0 & 0 & 0 & 0 & 0 & 0 & 1 & 0 & 0 & 0 & 0 \\
v_{16} & 0 & 0 & 0 & 0 & 0 & 0 & 0 & 0 & 0 & 0 & 0 & 0 & 0 & 0 & 0 & 1 & 0 \\
v_{17} & 0 & 0 & 0 & 0 & 0 & 0 & 0 & 0 & 0 & 0 & 0 & 0 & 0 & 0 & 1 & 0 & 1 \\
\end{array}.
$$

2.2.5 Parallel Feeder Configuration

Fig. 2.10 shows an interconnection between two feeders fed by two different substations. In large distribution network areas where large loads must be served and a high degree of reliability is required, the parallel feeder configuration is often used. Under this arrangement, several utility services are set up in parallel configuration, creating a more reliable system. The advantage of the configuration is to allow a single transformer to fail but no extra loss of service to the part of the system [59]. The power in this system can be supplied with the secondary injection source through the tie switch at the distribution substation.

Note that if one transformer fails in a subsystem, the alternative transformer and its associated secondary network must carry the entire load and the directional overcurrent relays are supplied on the secondary switchgear [59]. The topology planning engineer must take care of the system design in size selection of the transformer as well as the secondary switchgear that can be doubled for this type of system to be effective. The implementation of the

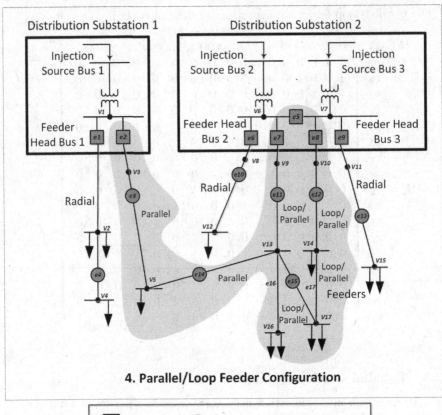

FIGURE 2.9
Parallel/loop feeder configuration.

incidence matrix for the parallel feeder configuration is illustrated as:

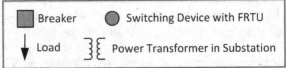

$$M_i = \begin{array}{c} \\ v_1 \\ v_2 \\ v_3 \\ v_4 \\ v_5 \\ v_6 \\ v_7 \\ v_8 \\ v_9 \\ v_{10} \\ v_{11} \end{array} \begin{pmatrix} e_1 & e_2 & e_3 & e_4 & e_5 & e_6 & e_7 & e_8 & e_9 & e_{10} \\ 1 & 1 & 0 & 0 & 0 & 0 & 0 & 0 & 0 & 0 \\ 1 & 0 & 1 & 0 & 0 & 0 & 0 & 0 & 0 & 0 \\ 0 & 1 & 0 & 1 & 0 & 0 & 0 & 0 & 0 & 0 \\ 0 & 0 & 1 & 0 & 0 & 0 & 0 & 0 & 0 & 0 \\ 0 & 0 & 0 & 1 & 1 & 0 & 0 & 0 & 0 & 1 \\ 0 & 0 & 0 & 0 & 1 & 0 & 0 & 0 & 0 & 0 \\ 0 & 0 & 0 & 0 & 1 & 1 & 1 & 0 & 0 & 0 \\ 0 & 0 & 0 & 0 & 1 & 1 & 0 & 0 & 1 & 0 \\ 0 & 0 & 0 & 0 & 0 & 0 & 1 & 1 & 0 & 0 \\ 0 & 0 & 0 & 0 & 0 & 0 & 0 & 1 & 0 & 0 \\ 0 & 0 & 0 & 0 & 0 & 0 & 0 & 0 & 1 & 1 \end{pmatrix}.$$

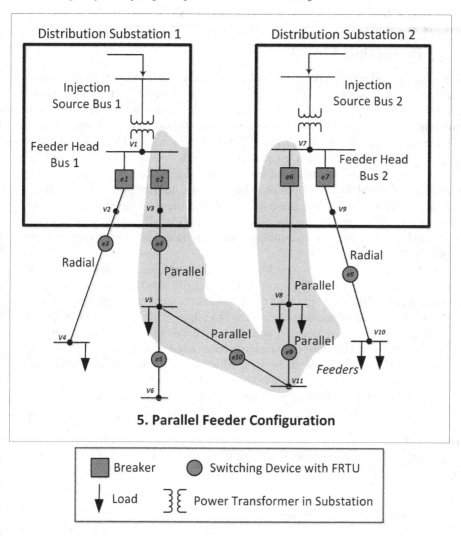

FIGURE 2.10
Parallel feeder configuration.

2.3   Reduction Model

Initially, the reduction model in graph theory is applied to simplify the working process and avoid duplicated operations by eliminating the superfluous procedures [60–62]. These procedures can be a block in a flowchart or vertices and edges in a graph. Similarly, the power flow applications are performed on the equivalent graph representation of a distribution network that is generally quite complex and involves a large number of edges and vertices. According to the economic consideration and the real geographical conditions, in the deployment of a real feeder, there may be multiple distribution nodes without any loads. These nodes are treated as the redundant nodes and may cause repetitive calculations during the power flow analysis. The reduction function is utilized to eliminate the redundant nodes and generate a relatively explicit topology.

Algorithm 3 shows the pseudocode of the topology reduction function based on the matrix form. The input of this algorithm is the incidence matrix. In this process, the rows' number of the input matrix represents the vertices sequence while the columns are served as the edges. This algorithm eliminates the redundant nodes by deleting the rows with the sum value as 2 since the redundant node in this case only has one enter edge and one leave edge. Then it combines the edges as one if they shared at least one redundant node in the initial topology. This procedure also involves the topological reordering. The corresponding MATLAB programming code is illustrated in Listing 2.2.

Listing 2.2
Topology Reduction Function Based on the Matrix Form

```
1   % Topology Reduction Function Based on Incidence Matrix
2   % INPUT: IncM is the incidence matrix
3
4   function IncM=redfuc(IncM)
5   %sum of each rows
6   temp = sum(IncM,2);
7   %find the rows with sum value 2
8   idx = find(temp==2);
9   %delete these rows
10  IncM(idx,:)=[];
11  d = IncM;
12  %sum of each columns
13  tmp = sum(IncM,1);
14  %find the columns with sum value 0
15  ind = find(tmp==0);
16  his = [];
17  %sign value
18  T = -1;
19  for i = 1:length(ind)
20      if ind(i)==T
21          continue
```

Algorithm 3 Topology Reduction Function Algorithm

Require:
    The topology is a radial network;
    The input is an incidence matrix with optimal ordering;
    Set **M** as input incidence matrix;
    Set **T** as identify mark;
    Set **ti** as the left column of the all-0s column;
    Set **td** as the right column of the all-0s column;
Ensure: the incidence matrix **M** without redundant vertices
    Input matrix **M**;
    Calculate the sum of each of the rows;
    Find and delete the rows with sum value 2;
    Calculate the sum of each of the columns;
    Find the index **ind** of the columns with sum value 0;
    for $i \leftarrow 1 :$ **length(ind) do**
        $ti = ind(i) - 1$; %The left column of $ind(i)$.
        $td = ind(i) + 1$; %The right column of $ind(i)$.
        while $sum(M(:, ti)) = 0$ **do**
            $ti \leftarrow ti - 1$; %Change the index of the left column.
        end while
        while $sum(M(:, td)) = 0$ **do**
            $td \leftarrow td + 1$; %Change the index of the right column.
        end while
        $M(:, ti) \leftarrow M(:, ti) + M(:, td)$;
        %Connect adjacency vertices after deleting the redundant vertices between them.
        Delete the $M(:, td)$;
        Delete the all-0s columns;
    end for
    return M

```
22        end
23        %0 left column
24        ti = ind(i)-1;
25        while sum(IncM(:,ti))==0
26            ti = ti-1;
27        end
28        %0 right column
29        td = ind(i)+1;
30        while sum(IncM(:,td))==0
31            T = td;
32            td = td+1;
33        end
34        his = [his,td];
35        IncM(:,ti) = IncM(:,ti) + IncM(:,td);
36    end
37    ind = [ind,his];
```

```
38  IncM(:,ind)=[];
```

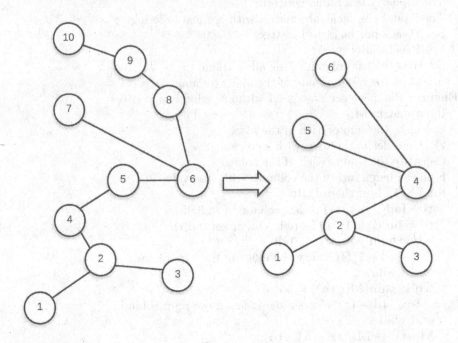

FIGURE 2.11
Topology change of the reduction function example.

Fig. 2.11 is an example of the reduction function. The initial topology consists of 10 nodes and 9 edges. The reduction function deletes nodes 4, 5, 8, and 9 and reorders the vertex index 6 as 4, 7 as 5, and 10 as 6. The corresponding incidence matrix of the initial topology is demonstrated as below:

$$
\begin{array}{c}
\\
v_1 \\ v_2 \\ v_3 \\ v_4 \\ v_5 \\ v_6 \\ v_7 \\ v_8 \\ v_9 \\ v_{10}
\end{array}
\begin{array}{c}
e_1\ \ e_2\ \ e_3\ \ e_4\ \ e_5\ \ e_6\ \ e_7\ \ e_8\ \ e_9 \\
\left(\begin{array}{ccccccccc}
1 & 0 & 0 & 0 & 0 & 0 & 0 & 0 & 0 \\
1 & 1 & 0 & 0 & 0 & 0 & 0 & 1 & 0 \\
0 & 0 & 0 & 0 & 0 & 0 & 0 & 1 & 0 \\
0 & 1 & 1 & 0 & 0 & 0 & 0 & 0 & 0 \\
0 & 0 & 1 & 1 & 0 & 0 & 0 & 0 & 0 \\
0 & 0 & 0 & 1 & 1 & 0 & 0 & 0 & 0 \\
0 & 0 & 0 & 0 & 0 & 0 & 0 & 0 & 1 \\
0 & 0 & 0 & 0 & 1 & 1 & 0 & 0 & 0 \\
0 & 0 & 0 & 0 & 0 & 1 & 1 & 0 & 0 \\
0 & 0 & 0 & 0 & 0 & 0 & 1 & 0 & 0
\end{array}\right)
\end{array}
$$

and it has been changed to:

$$
\begin{array}{c}
 \\
v_1 \\
v_2 \\
v_3 \\
v_4 \\
v_5 \\
v_6
\end{array}
\begin{array}{ccccc}
e_1 & e_2 & e_3 & e_4 & e_5 \\
\left(\begin{array}{ccccc}
1 & 0 & 0 & 0 & 0 \\
1 & 1 & 0 & 1 & 0 \\
0 & 0 & 0 & 1 & 0 \\
0 & 1 & 1 & 0 & 1 \\
0 & 0 & 0 & 0 & 1 \\
0 & 0 & 1 & 0 & 0
\end{array}\right)
\end{array}.
$$

Mini Project 1: Distribution Topology Reduction

The five configuration cases introduced in this project provide the combination of possible topological statuses. Create more fictive nodes in the existing test cases and execute the scripts to verify the reduced incidence matrix. Justify whether this process to add new nodes is efficient. If not, what would you suggest to improve the efficiency of this process. (Hints: the promising solution to topology maintenance is Chapter 3.)

2.4 Conclusions

This chapter establishes the fundamental language of distribution network elements in graph modeling. The formation of a matrix is established and the traverse algorithms are described. As the feeders are radially energized under normal operating conditions, the exploration of nodes ensures the unique establishment of various types in distribution network topology during the transition from emergency back to normal. The "transient" state of topology is to infer potential segments within a feeder to pinpoint a faulted location. This chapter also discusses the reduction of the network model, as not every (fictive) node of line terminals connects to a load/capacitor.

This chapter also introduces the importance of reordering and reduction of a graph. Reordering can be based on a feeder that would organize the incidence matrix to be intuitive visually. The reduction ensures additional nodes that are not connected to any loads or capacitors (both are one-terminal elements). In the future chapter, reduction may occur to simplify the graph by adding an equivalent lump load between two immediate switches (typically remote-controlled switches). Extraction of geospatial datasets is introduced in the next chapter. The formation of a graph introduced in this chapter can be directly extracted from the geographic information system (GIS) database where it is commonly used in the planning department of a utility.

3

Geospatial and Topological Data Establishment

In the management and operation of electrical distribution systems, the topology construction is required since the dynamic geographical schematics need to be shown in the SCADA system [63]. In most cases, the topology of an electrical distribution network is updated regularly. These constant updates are maintained by the planning engineers in a distribution company so as have bookkeeping of the latest status of topological updates in terms of adding/removing/moving distribution transformers, new line segments, or switches within the network.

The GIS [64] is used to maintain the incremental update of these datasets to ensure the integrity of the database is up-to-date. The schematic diagrams available in GIS generally cannot be used in the same format as in the DMS for monitoring and controlling the distribution network. Therefore, SCADA systems require additional picture data to describe their graphical displays and physical topology. These datasets describe all diagram objects in terms of captions, measurement values, and how these datasets should be displayed on the map as well as the symbol or sketch. Static data and object data must be linked to the same SCADA device or element associated with the distribution node data, which is dynamically reflected by the system energization states.

In addition, the template for the SCADA CIM format would conform to the topological connectivity to link with additional measurements of the RTU/FRTU [65,66]. These inputs are part of the importation to the SCADA systems described in the previous chapters. These data inputs should be systematically organized with coordination of the planning department and operational department to verify the consistency as part of the data engineering process. The complexity may vary, depending on the proprietary modeling of an organization, but this importation serves as the process to automate the process for the first time and incremental updates over time. The inconsistencies with errors to SCADA datasets can be pre-processed first on the topological verification before them importing to the SCADA systems [67]. The data exchange between GIS and SCADA is the typical routine of planning engineers to ensure the models are validated and verified [68,69]. Essentially, these topological datasets, if correctly improved over time, would have values in applications to be used in DMS applications such as power flow, outage management, etc., to be discussed in later chapters.

3.1   Understanding the GIS Topology

GIS is a data visualization system to display the geographical or topological information from the real earth's surface. The GIS database is constructed based on a real map. The database can be used to label specific position information, capture related data from a topology, classify different data types, and store and extract information for analysis [70]. GIS can help designers or operators better understand spatial patterns and relationships of a system topology by relating seemingly unrelated geographical data [71, 72].

The layers of GIS datasets can be generalized into two types: (1) electrical layers and (2) non-electrical layers. Fig. 3.1 is the geographical demonstration of a distribution network. The illustrated network is constructed of multiple static layers (road, building, territory, etc.) and dynamic layers (distribution nodes, loads, or switches). The geography of locations that depicts the lines and points can be the elements of an electrical feeder. The main feature of overlaying multiple layers of information under the 2-dimensional map is to provide an intuitive view of a network for the purpose of assimilation between these layers. It is at the organization's discretion what layers should be displayed in the real-time map and what layers should be used to correlate with other information. In some cases, a 3-dimensional view of a map could include additional information that would provide insight based on its causality, i.e., display risks and potential impacts.

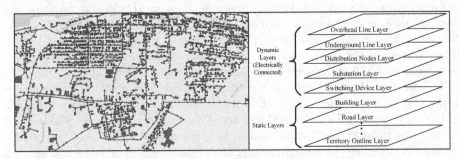

FIGURE 3.1
Geographical representation of a distribution network in multiple layers.

3.1.1   Preparing Electrical Layers

A geographical map may consist of layers that are based on GIS datasets. These datasets can be represented in the form of points, polylines, or polygons. The representation of all these entities may be topologically related to each other. We refer this to our earlier discussion on electrical layers as an integral part of the electrical networks. The non-electrical layers may be used for display purposes to correlate with the electrical layers. However, the topology of non-electrical layers may be useful, such as a road map that may be used for a crew to travel. This can be provided by other applications that are built on the existing SCADA/DMS applications. The format of GIS files can also vary. Typically, environmental systems research institute (ESRI) data is stored in geodatabases, coverages, shapefiles, imagery, rasters, or CAD files. This data is embedded with topology and other information related to each element. It would be unusual to have a format that is stored in the ASCII format about the topological connectedness. The conversion between shapefile and database is shown in Fig. 3.2.

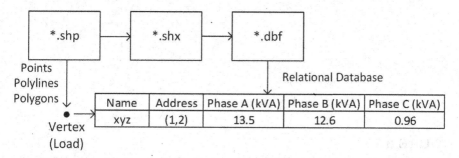

FIGURE 3.2
The conversion between shapefile and database.

3.1.1.1 Basemap

The basemap is input from the real map gallery or by using the self-designed layer. It can provide the geographical background for a topology or network to display. The basemap can be changed in the current map at any time. Fig. 3.5 shows a comparison between a GIS map with a different basemap.

3.1.1.2 Points

Points are the simplest features to create. The point can be added to a new points template or to indicate the end of a line. Different points layers can represent different themes; as shown in Fig. 3.3, two different points layers denote the distribution substation and nodes separately.

Populate all geometry nodes:

- Write a function to repeatedly use all layers and determine the total number of the nodes in a given GIS set.

- Determine the unique list of the total nodes. Write two functions to handle single- or two-terminal elements separately.

  - For single-terminal elements: go through all of the layers.

- For two-terminal elements: go through the heads and tails of each element. Combine the two lists of heads and tails into one list.

- Assign a unique number to each node (pairing function).

- Compare the total elements of the original list and unique list.

- Check topological errors.

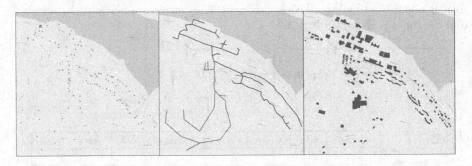

FIGURE 3.3
Campus-wide distribution system points layer, polyline layer, and polygon layer in GIS.

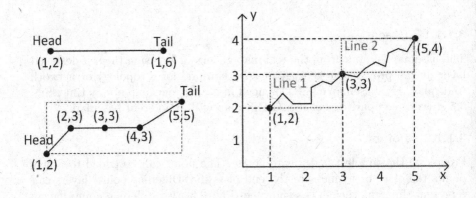

FIGURE 3.4
Polylines in GIS.

3.1.1.3 Polylines

The polyline in GIS, as shown in Fig. 3.4, can be created by a finished 2-point single-part polyline feature, a radial feature, or a free-form polyline feature using the pointer. In addition, new polylines can be drawn to snap the existing lines together. In Fig. 3.3, two different polyline layers represent the overhead and underground lines in a distribution system separately. All polylines in this figure are created based on the existing nodes.

3.1.1.4 Polygons

Different than polylines, the tail (end point) and the head (start point) end on the same xy-coordinate in polygons. In addition, a polygon is created based on the coordinates defined in the coordinates area or according to the existing basemap. It is the first step to create a new geometry or it can be treated as part (the planar structure) of a more complex geometry. The polygon can also be drawn through freehand. Fig. 3.3 demonstrates "building" polygons in the example topology. Fig. 3.5 shows the integrated geographical map of the campus-wide distribution network with five layers.

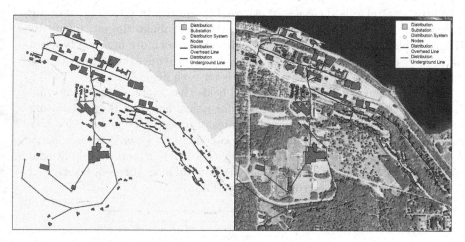

FIGURE 3.5
Campus-wide distribution system with different basemap in GIS.

3.2 GIS Data Extraction

Data extraction is the process of retrieving the data of interest and/or requirements from the original data sources and converting the relevant data in a specific pattern for further data processing or data storage. The original data sources are usually unstructured or poorly structured. The majority of data extraction is to identify the diverse data formats.

The extraction of the geographic data from GIS is based on the data attribute values, the spatial range, and the geographical features. Using the extraction tools in GIS, the attributes information of the selected layers in a specified area can be extracted to a required format with the spatial reference. A new subset of attributes or features can be generated for analysis.

MATLAB functions for GIS data extraction:

- shaperead: read all shape files into memroy.

- unique: determination of a unique set.

- sparse: conversion from MATLAB array into sparse format.

- spy: visualize the sparsity of a matrix.

3.2.1   Attribute Table in GIS

The attribute table is the basis of geographic features in GIS. The data are visualizable, queryable, and analyzable within the tabular format. In the attribute table, columns represent the fields, which can store multiple types of data, such as the number ID, capturing date, or a brief description (text), while rows store the corresponding records for each field [73].

Feature classes are tables with special fields that contain related information about the geometry of shape fields for point, line, and polygon feature classes. Some fields, such as the coordinates of objects, the unique identifier number (ObjectID), and Shape (point, polyline, or polygon), are automatically added, populated, and maintained by GIS [73]. Furthermore, additional fields and features can be added to the attribute table for different layers. Fig. 3.6 is an example of the attributes table in GIS.

FID	Shape *	X	Y
0	Point	-9857509.93895	5961877.1553
1	Point	-9857392.27119	5961888.48254
2	Point	-9857362.53863	5961970.49663
3	Point	-9857174.68408	5961909.64235
4	Point	-9856976.24619	5961729.72532
5	Point	-9856947.95338	5961791.66625
6	Point	-9856908.8308	5961646.8832
7	Point	-9857475.89028	5961784.50186
8	Point	-9857536.74459	5961792.43948
9	Point	-9857601.94672	5961760.11957
10	Point	-9857623.1134	5961728.36963
11	Point	-9857470.59865	5961676.02249
12	Point	-9857340.00867	5961614.59839
13	Point	-9857250.99397	5961580.77243
14	Point	-9857181.88671	5961548.45929
15	Point	-9857185.96885	5961499.87224
16	Point	-9857041.65727	5961508.77171
17	Point	-9856869.04941	5961458.49388
18	Point	-9856826.71599	5961543.16072
19	Point	-9856882.86575	5961544.01575
20	Point	-9856787.02841	5961604.015

FIGURE 3.6
Attribute table of points layer in the campus-wide topology.

The attribute table can export as the Excel file, .DBF, .txt, etc.

3.2.2   Pairing Function

The pairing function is applied to uniquely encode two natural numbers into a single natural number [74]. In the conversion between graph or GIS shapefile to the matrix format, this function is needed to concatenate the xy coordinate from the .shp file into a number for each element.

Mathematically, let $z \in \mathcal{N}$ be an arbitrary natural number. There exist unique values $x, y \in \mathcal{N}$ that:

$$z = \pi(x, y) = \frac{(x + y + 1)(x + y)}{2} + y.$$

The pairing function is shown in MATLAB programming as Listing 3.1.

Listing 3.1
Pairing Function Example

```
1  % In mathematics a pairing function is a process to uniquely ...
       encode two
2  % natural numbers into a single natural number. Any pairing ...
       function can
3  % be used in set theory to prove that integers and rational ...
       numbers have
4  % the same cardinality as natural numbers.
5
6  function [ pair_output ] = pairfunc( x, y )
7      pair_output = 1/2 * (x + y) * (x + y + 1) + y;
8  end
```

3.2.3   Graph-to-Matrix Conversion Algorithm

The adjacency matrix is a square $|V| \times |V|$ symmetric matrix that reflects the adjacencies between vertices in a graph, while the incidence matrix is a $|V| \times |E|$ matrix that displays vertices' and edges' incidences. These two matrices can be converted to each other. Since the adjacency or incidence matrix can display the straightforward interrelationship between vertices and edges, the matrix is employed to store and fix the current connection state of the network topology. Furthermore, this process converts the topology to the mathematic model, which is machine recognizable. The further relevant algorithm and analysis can be performed based on the matrix [50].

Algorithm 4 is the pseudocode that shows how the GIS schematic diagram converts to the adjacency matrix. This algorithm can be used not only for a simple virtual network but also for a large and complex geographical related topology. The input of this algorithm requires a shapefile. The algorithm is summarized as follows:

- Require the shapefile of the GIS diagram $G = (V, E)$, where V is the set of vertices and E is the set of edges;

- Create a pairing function f_{pair} that can uniquely encode two natural numbers into a single natural number [75, 76]. This function will take the x and y coordinate individually and map it into a unique number that will not overlap with other xy-coordinate combinations;

- Make each edge's start vertex that is close to the power source be the head node and the end vertex be the tail node;

- Perform f_{pair} for each of the edge head and tail nodes;

- List all head and tail nodes, eliminate the repeated elements;

- Renew the head and tail datasets with corresponding nodes indexes;

- Initialize an empty adjacency matrix;

- Check each edge in the graph. If the head index is not equal to the tail index, indicate 1 as a connected state to the corresponding position in the adjacency matrix;

- After generating the adjacency matrix, convert it to the incidence matrix with the existing method [51, 77];

- Return the adjacency and incidence matrices.

The MATLAB demonstration of the GIS topology-to-matrix conversion algorithm is shown as Listing 3.2. The inputs of this script are shapefiles from GIS.

The function "extract1ter" is utilized to read shapefiles and apply the pairing function to take the x and y coordinate individually and map it into a unique number that will not overlap with other xy-coordinates. The "extract1ter" function is illustrated in Listing 3.3.

Listing 3.2
GIS Topology to Matrix Conversion

```
1  % Read all 1ter and 2ters in shapefile format
2  Load=extract1ter('Improved 2 - delete cycle lines\Load');
3  Switch=extract1ter('Improved 2 - delete cycle lines\Recloser');
4  Cap=extract1ter('Improved 2 - delete cycle lines\capacitor');
5  [head, tail] = extractheadtail('Improved 2 - delete cycle ...
       lines\aclines');
6  headSize = size(head,1)
7  tailSize = size(tail,1)
8
9  Line2Ters    = [head; tail];
10 Line2TersLoad = [Line2Ters; Load];
11 Line2TersLoadCap = [Line2Ters; Load; Cap];
```

Algorithm 4 Graph-to-Matrix Conversion Algorithm

Input:
   $G = (V, E)$
Output:
   M_a, M_i
1: **for** each edge **do**
2:    x_{head} = x-coordinate of edge's start vertex;
3:    y_{head} = y-coordinate of edge's start vertex;
4:    x_{tail} = x-coordinate of edge's end vertex;
5:    y_{tail} = y-coordinate of edge's end vertex.
6:    Define the pairing function f_{pair}.
7:    Head = $f_{\text{pair}}(x_{\text{head}}, y_{\text{head}})$
8:    Tail = $f_{\text{pair}}(x_{\text{tail}}, y_{\text{tail}})$
9: **end for**
10: Node = unique ([Head; Tail]).
11: Head = find (Head = Node).
12: Tail = find (Tail = Node).
13: % Initialization. Make sure this is a square matrix.
14: M_a = zeros (length(Head), length(Head)).
15: **for** i = 1:length(Head) **do**
16:    **if** Head(i) $\neq$ Tail(i) **then**
17:       M_a (Head(i), Tail(i)) = 1.
18:       M_a (Tail(i), Head(i)) = 1.
19:    **else**
20:       M_a (Head(i), Tail(i)) = 0.
21:    **end if**
22: **end for**
23: Convert M_a to M_i [51, 77].
24: **return** M_a, M_i.

```
12  Line2TersLoadCapSw = [Line2Ters; Load; Cap; Switch];
13
14  UniNodes = unique(Line2Ters); %column vector
15  head = IndexElem(UniNodes, head);
16  tail = IndexElem(UniNodes, tail);
17
18  UniNodesAll = unique(Line2TersLoadCapSw); %column vector
19  Load = IndexElem(UniNodesAll, Load);
20  Switch = IndexElem(UniNodesAll, Switch);
21  Cap = IndexElem(UniNodesAll, Cap);
22
23  Y=zeros(headSize,headSize); % Make sure this is a square matrix
24  jj=1;
25  for ii = 1:size(head,1)
26      if (head(ii,2)≠tail(ii,2))
27          Y(head(ii,2),tail(ii,2))=1;
28          %Y(tail(ii,2),head(ii,2))=1;
```

```
29        else
30            ErrorV(jj,1)=ii;
31            Y(head(ii,2),tail(ii,2))=0; % mark as non-existent ...
                 nodes due to GIS shapefile topological error
32            jj=jj+1;
33        end
34    end
35    spy(Y)
36
37    N=size(Y,1)
38    A=Y(1:N, 1:N);  % Adjacency matrix
39    Adj= sparse(A); % successfully convert into sparse format
40    Inc= adj2inc(A); % incidence matrix
41
42    [xx,yy]=find(Adj(:,:)==1)
43    G = sparse(xx',yy',true,N,N)
44
45    check=graphisdag(G) % check if this is a directed acyclic graph
46    if check
47        order = topological_order(G)
48        G = G(order,order)
49        spy(G)
50    else
51        cyclicloop=findcycles(G)
52    end
53
54    bg=biograph(G(1:300,1:300),[],'ShowArrows','off')
55    view(bg)
56    %A=graphminspantree(G)
```

Listing 3.3
Extraction of One-Terminal Element

```
1    function [ xy ] = extract1ter(filename)
2        [Shape, Database] = shaperead(filename); % shape file read
3        N1Terminal = size(Shape,1);
4        for ii = 1: N1Terminal
5            xy(ii,1)=pairfunc(Shape(ii).X,Shape(ii).Y); % recall ...
                 pairing function
6        end
7    end
```

The function "extractheadtail" is applied to identify the head and tail nodes for a two-terminal element (edge or link), and the function "IndexElem" is used to filter the indices of "heads" and "tails" as the unique number. The MATLAB scripts of these two functions are demonstrated in Listings 3.4 and 3.5.

Listing 3.4
Extraction of Head and Tail for a Two-Terminal Element

```
1    function [ AHead, ATail ] = extractheadtail(filename)
```

```
2      [shape, database] = shaperead(filename);
3      N2Terminals=size(shape,1)
4      for ii=1:N2Terminals
5          % The last record is actually before NaN
6          NArraySize=size(shape(ii).X,2)-1;
7
8          % Determine xy-coordinate for a head of a two-terminal ...
                element
9          headX = shape(ii).X(1);
10         headY = shape(ii).Y(1);
11         AHead(ii,1)=pairfunc(headX,headY);
12
13         % Determine xy-coordinate for a tail of a two-terminal ...
                element
14         tailX = shape(ii).X(NArraySize);
15         tailY = shape(ii).Y(NArraySize);
16         ATail(ii,1)=pairfunc(tailX,tailY);
17     end
18  end
```

Listing 3.5
Identify the Unique Index for Each Head and Tail

```
1  function [ TermElem ] = IndexElem(UniNodes, TermElem)
2
3  for ii = 1 : size(TermElem,1)
4      TermElem(ii,2)=find(UniNodes==TermElem(ii,1));
5  end
6
7  end
```

Use the distribution network shown in Fig. 3.5 as an example. The sparse format of the converted incidence and adjacency matrices are shown in Fig. 3.7.

Another complex example is demonstrated in Fig. 3.8, which is a geographical region of the distribution test network. There are 3 substations, 1,665 distribution nodes, and 1,644 buildings (including residential and non-residential). The sparse format of the converted adjacency and incidence matrices is shown in Fig. 3.9.

3.3 Conclusions

Each feeder is connected with thousands of different elements. The electrically connected establishment will be useful in later chapters to pinpoint or infer potential faults. This chapter provides a means of topology access and introduces how extraction of topology can be established. Electrical and non-electrical layers of the GIS datasets are described. The electrical layers include points

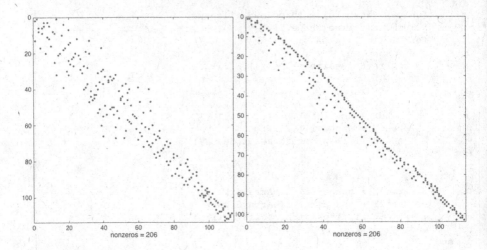

FIGURE 3.7

Sparsity visualization of the adjacency and incidence matrices for the campus-wide distribution network.

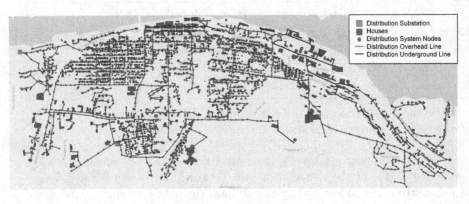

FIGURE 3.8

The test distribution system in geographical proximity.

and polylines that describe electrically connected elements of a (sub-)system. The pairing function described in this chapter ensures the uniqueness of xy-coordinates and the connectivity of the described feeders. As the distribution network topology evolves over time, the GIS serves as the reference point of database maintenance so that planning engineers can keep this layer updated. Chapters 2 and 3 are the foundation of network topology, where many applications will rely on it. Like the notion of "garbage in, garbage out" by W. H. Kersting, it is critical to maintain a close-to-real-world status that reflects a current network topology.

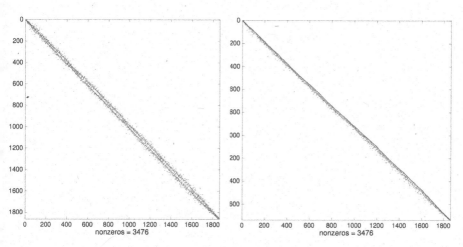

FIGURE 3.9
Sparsity visualization of the adjacency and incidence matrices for the proposed topology.

The GIS topology introduced in this chapter provides the original version of a given graph that can be simplified. In future chapters, the topology can be simplified only between switches (remote-controlled and non-remote-controlled switches) where each area between switches is simplified with a lumped load between switches, indicating the number of associated customers and equivalent impedance information. The most simplified version of the topology is to only include remote-controlled switches. However, this visual representation or graph will lose significant information with regard to the number of non-remote-controlled switches in between, which can affect a large number of customers when the faulted area is isolated by two remote-controlled switches. That is why both types of switches are included in Chapter 8, in order to narrow down the faulted area.

Mini Project 2: GIS Datasets

This chapter introduces the relational database (**dbf** file) with the geometry file (**shp** file) of the GIS system. Using ESRI ArcGIS, create 5 layers that describe the electrical layers, i.e., loads, lines, capacitor, switches, and distribution transformers. Make sure the connection of the line terminals are snapped to the other point/polyline layers to ensure connectedness. Describe whether each of them is consistent in distribution system modeling and the shape file format. For instance, a load is a point layer and also consists of one terminal in the modeling. The switches are in the point layer but have two terminals in the graph modeling.

Part III

Computerized Management and Basic Network Applications

4

Unbalanced Three-Phase Distribution Power Flow

Advanced distribution management systems (DMS) become an extension of SCADA features by applying the software management principles from transmission networks to distribution systems. In addition, the modern DMS is capable of assessing the operational state of the distribution system and evaluating control functions such as reactive dispatch, voltage regulation, "what-if" scenarios, and analysis [78]. Real-time evaluation using a power flow module will provide a close-to-real-world scenario simulation, to predict the best consequences that can be useful for operational purposes [79]. A calculation or analysis method should be developed to meet the requirements of rigorous operational type analyses, system power losses, and hypothetical analyses on a large-scale distribution system that connects multiple substations.

Establishing an unbalanced power flow module is essential. This is because the program will play an important role to determine the unknown variables, such as voltage and angle for all load nodes, and reactive power and angle for the generator nodes. Upon a convergence of power flow calculation, the solution assures complete balance of a feeder. This is a very useful solution to determine the total real and reactive losses at a given time, in which hypothetical scenarios can be simulated,and the solutions with operational constraints [80,81] will provide a choice so that dispatcher in the control center can decide which tie switch they have to close to supply power to the "healthy" part of the area.

Visual representation of topology from a geographical map is essential. Fig. 4.1 shows a sample GIS feeder that consists of detailed geographical information. This chapter introduces how to extract topological information, shown as part (a). In graph modeling, part (b) shows the connectivity between lines, nodes, loads, and the switches (remote-controllable only). Part (c) is a simplified one-line diagram with remote-controllable switches (RCSs) and the lump load in between them. Part (c) will be used later to illustrate a specific area of interest for fault location and partial restoration. The lump load between the RCSs is useful because it can show intuitively the connection of RCSs for the entire feeder. Part (d) is similar to part (c), but includes the non-RCSs. Typically, between RCSs, there are a few non-RCSs between boundary RCS in the area.

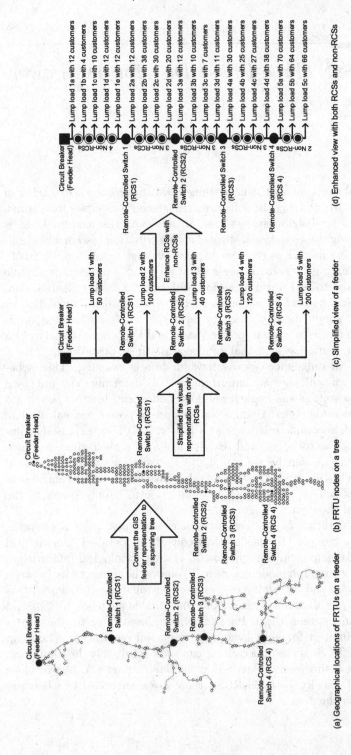

FIGURE 4.1

Sample feeder from geographic information system (GIS) to two one-line diagram representations (the left two).

4.1 Important Roles of Unbalanced Three-Phase Distribution Power Flow

Two basic ingredients needed to have the best outcome of power flow results are the accurate topology data and real-time measurement where the topology data is from GIS. In distribution systems, the most difficult part is the topology update where some errors incurred, topologically may cause power flow fails to converge [82]. As the measurements are either top-down, that is, pro-rated by allocation factors, or through load profile by the representative of load types, none of these are accurate enough to represent the close-to-real-time state due to the estimation. Real-time measurements from RTUs and FRTUs within a feeder can serve as the reliable live information to be used to project good estimates for each child node (loads) of a feeder.

The decision to make an investment in improving system observability is critical. The tree structure (radial) of a feeder would be an ideally good setup of observation if the "parent" and "children" nodes have meters deployed. However, this is not economically feasible because the ongoing costs of maintenance and emergency conditions are often uncertain. As the system is evolved toward more automation and more sensors around primary and secondary networks, the numerical analysis for power flow calculation is becoming very burdensome. For example, some nodes are often estimated based on offline information, which can be far away from the real state of the system in general [83].

Since the practical design of overhead lines, cables, and transformers in normal conditions has to meet the requirements of a particular situation, the real distribution systems are inherently unbalanced [84]. The asymmetrical phase spacing, the line segments with different phases (single-, double- and three-phase), and the imbalance loads will cause an unbalanced distribution system [85].

Distribution systems are typically not balanced due to the inconsistent loads connected to the feeders and laterals. When a short circuit analysis is assumed in a system, the unbalanced conditions can be categorized into two scenarios: short-term unbalances and steady-state unbalances [84]. Short-term unbalances include single-phase fault, double-phase fault, double-phase fault, and other combinations of faults from the 3-phase circuits. Reclosers and breakers feature auto-reclosure that will lock out after several attempts (typical 3 times). Such phenomena produce a temporary balance as well, which does not last long. The characterization of this type of fault is dynamic, such that it causes temporary overloads that are often very short.

The second category is a permanent condition of a fault; the recloser or breaker locks out after several attempts of reclosing that confirm the scenario requires special attention to troubleshoot it. The combinations of scenarios include the following [84]:

- damaged transmission lines,

- one or two phases malfunction in a breaker,

- unbalanced loads,

- different single-phase units in one transformer.

Due to the unbalance of the system and the unbalanced nature of faults, itemized three-phase models must be employed in order to accurately study the scenarios for operating analysis. To handle such detailed models, the three-phase power flow approach is necessary to perform computational analyses related to the distribution system operation.

4.2 Calculation Methods about Power Flow Analysis

Conventional power flow methods applied in the transmission and sub-transmission systems have been well developed. However, since the distribution system is quite different in some aspects from a transmission system, the conventional approaches are inefficient in dealing with the analysis in distribution networks. The distribution system keeps a radial topology or a temporary weakly meshed network in contrast with transmission topology that is "tightly" meshed [86]. The distribution system is a low voltage system with a wide range of reactance and resistance values. The distribution system could consist of a large number of clusters or sections, such as the microgrids, and fictive nodes spread throughout the whole network. Because of the unbalanced three-phase load as well as the existence of single- and double-phase loads in a lumped load area, there could be unbalanced loads in the distributed sections of the system. Since there are rare transposed lines in a distribution system, the mutual couplings between phases cannot be neglected [87, 88].

The applicable solution techniques of power flow are classified according to the network topology and the adopted reference frame. For a loop topology or a general network structure, the power flow analysis is conducted based on Newton–Raphson and Gauss and/or Gauss–Seidel methods. As expected, the Newton–Raphson-based approach (including fast-decoupled versions) is well adapted to the mesh topology such as in a power transmission network that converges very often. However, the network configuration for a distribution system is the radial configuration, which may require special treatment on R/X ratio on the feeders and laterals. This can cause a singularity problem, leading to a diverged solution [85, 89]. The phase-frame approach, the forward-backward sweep, and the compensation approach have been presented for power flow analysis in radial topology and weakly meshed networks, respectively [90].

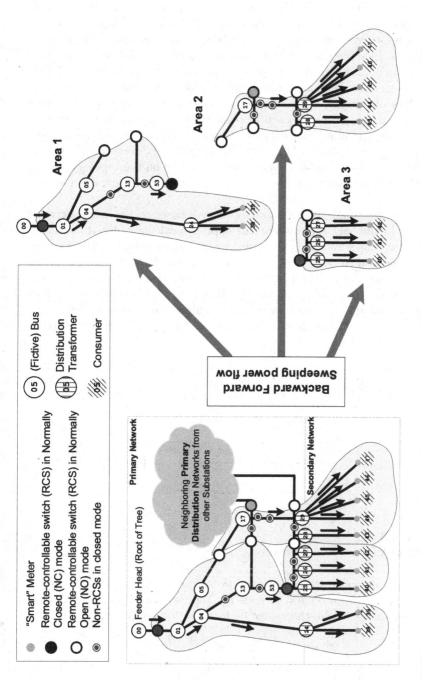

FIGURE 4.2
Division of 3 areas based on the FRTU root nodes in a feeder [91].

Fig. 4.2 shows an example feeder with 3 NC RCSs and 4 NO RCSs. Technically, the power flow module may handle and evaluate the entire feeder as a system (feeder by feeder). However, each of the NC RCSs has the 3-phase power flow information from the FRTUs where it can be performed separately for each area. In some cases where there is a topology error in one of the 3 areas, division of 3 areas of running power flow can assure convergence reliability of solutions in some areas. As each distribution transformer (DT) requires initial values of current, the allocation factor is used to appropriate active power and reactive power for each associated DT from the immediate root node (the metered value from FRTU) of an area. Typically, the power flow module is executed based on (1) a periodic trigger, e.g., every 10 minutes or (2) event-driven. The event-driven execution is based on the topology status, such as the recloser under a feeder is tripped or when operators open an NC RCS.

The power flow analysis in a radial topology is based on the ladder network theory [80]. In the backward sweep, this approach calculates the distribution system bus voltages and adds the currents for each section. In the forward sweep, this method corrects the substation voltage on the feeder head with the initialized voltage value. The bus voltages in the system are calculated again in this iteration. The topology during this backward and forward process does not change, and the approach meets the requirements of circuit laws. The convergence process will consist of multiple iterations, and all bus voltages are computed twice in each iteration.

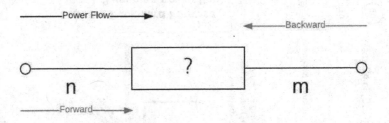

FIGURE 4.3
Schematic of forward/backward sweep.

Figure 4.3 demonstrates the schematic of the forward/backward sweep. The power flow is from node n to node m. The forward sweep computes the downstream voltages from the source (substation) by applying [41]:

$$[V_{abc,m}] = [A] \cdot [V_{abc,n}] - [B] \cdot [I_{abc}].$$

To start the process, the load currents $[I_{abc}]$ are initialized as zero and the voltages are calculated. In the first iteration, the load voltages are the same as the source voltages.

The backward sweep computes the currents from the load back to the source using the most recently computed voltages from the forward sweep as [41]:

$$[I_{abc,n}] = [c] \cdot [V_{abc,m}] + [d] \cdot [I_{abc,m}],$$

and the bus m voltage can also represented as:

$$[V_{abc,m}] = [a]^{-1} \cdot [V_{abc,n}] - [b] \cdot [I_{abc}],$$

where $[A] = [a]^{-1}$ and $[B] = [a]^{-1} \cdot [b]$.

In the case here, the a, b, c, d, A, and B are 3×3 generated matrices and will be referred to as the generalized line matrices [41].

After the first forward and backward sweeps, the updated load voltages are calculated based on the most recent currents. The forward/backward sweep continues until the mismatch of the load voltages that results between the new and the previous iterations is within the specified tolerance.

Another schematic is shown in Fig. 4.4, with different levels. Nodes 3, 4, and 5 belong to level 3. Then, the node sequence of the backward step is $[6, 4, 5, 3, 2, 1]$ while the level sequence is $[4, 3, 2, 1]$. For the forward step, the sequences are $[1, 2, 3, 5, 4, 6]$ and $[1, 2, 3, 4]$.

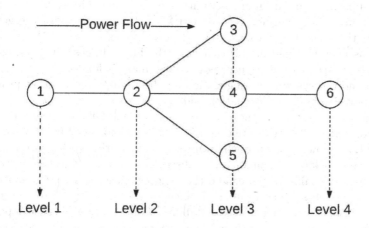

FIGURE 4.4
Forward/backward sweep with different levels.

The IEEE 4 Node Test Feeder is shown in Fig. 4.5. The source line segment is from node 1 to node 2 and the load line segment is from node 3 to node 4. The flow chart shown in Fig. 4.6 illustrates the forward and backward sweeps based on this distribution feeder. The corresponding indexes and matrices are shown in the figures.

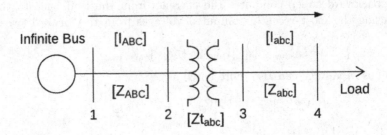

FIGURE 4.5
Example to forward/backward sweep [41].

4.3 Power Flow Analysis for Feeder Switching Operations

Topological reconfiguration by switching procedures in the distribution system is an important operation of distribution automation. The system reconfiguration is usually performed to isolate a faulted area or achieve the goal to optimize the system operations and reduce power losses. Since the whole topology has been changed in this situation, the bus voltages, currents, and losses will be re-distributed. Hence, power flow analysis for feeder switching operations and the system reconfiguration has been employed to solve problems incorporating system automation and operations.

Overloads do occur sometimes to stress the system. This means distribution transformers under stress are expected to have a shortened life span over time. Supplying power via an open transformer connection to 3-phase loads may even cause a larger imbalance in a distribution network. When a fault occurs on the first half of a feeder and the recloser (closer to it) trips, half of the feeder experiences a power outage. If a feeder has no tie switch to another substation feeder, this can hurt the reliability of the system because repair time can be significant. If there is a NO tie switch connected to an adjacent feeder from the downstream, the crew can find out the minimum area in which to isolate the faulted area by opening those switches while it is still not energized. Then, the healthy part of the network will be temporarily energized from the tie switch of the downstream distribution network.

The combinations of NO switches give the dispatchers a choice of the best option among them while maintaining the radial network during emergency conditions. A faulted feeder with more than one NO switch can be simulated using a power flow module. Connecting the neighboring feeders can be simulated in parallel as the slack bus of each feeder may not be the same substation. These NO switches are often connected with FRTUs (which will be

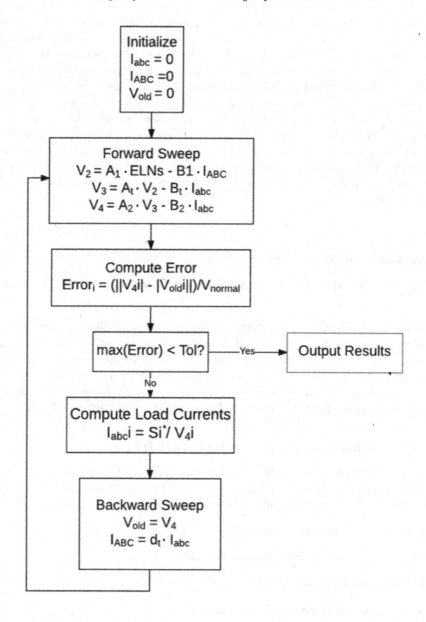

FIGURE 4.6
Flow chart of forward/backward sweep based on the example [41].

discussed in the next section), which represents an upgrade from the section-alizing switches in a radial feeder. These remote-controlled switches can be in either closed or open states. A full-blown DMS system requires a power flow module as the base to determine the scenarios based on the latest telemetered switches' information to evaluate and recommend options, i.e., which switches should be open and which switches should be closed. The reconfiguration can be a simple switching state for any switch that is involved in the power outage area. A power flow module provides the best options by considering the system's operating constraints at a given time. Other constraints may include transformer loading, feeder thermal loading, feeder current balance, voltage-drop, and line losses [92, 93]. A feeder reconfiguration would depend on the available remote-controlled switches that are connected with other feeders. Together with the latest operating condition, the topological statuses provide the power flow modules to find out the best options that dispatcher with a way can rely on.

4.4 Available Metering Sources within Distribution Feeders

All DMS applications rely on the availability and reliability of the SCADA meter measurements. The power flow module is no exception. The injection points (i.e., each feeder head) for power flow analysis are often attached to the analog measurement variables to initialize the backward/forward sweeping unbalanced power flow. This section describes the available metering infrastructure from the primary and secondary parts of a distribution system.

4.4.1 Measurements from FRTU and RTU

The SCADA system is capable of monitoring the operation status of the system, managing the system database, and meeting the requirements of system operations. The main components of a SCADA system consist of

- the master computer station and control panel,
- communication media (wired and wireless),
- remote terminal unit (RTU),
- pole-mounted feeder remote terminal unit (FRTU),
- human-machine interface (HMI),
- and field equipment, etc. [94–96].

As the main components of SCADA, the main purpose of RTUs/FRTUs is to capture relevant information from the system and send it to the higher hierarchy to analysis, or to show the data to the operators. Multiported RTUs instantaneously provide data to the regional master stations, management center, and/or neighboring utilities without the time delay [94, 97].

The instrumentation is where the devices report the state of the local measurement to the SCADA system. There are two forms of data measurements: binary status and analog values. Binary status is the quantities that are obtained by circuit breakers, pole-mounted switches, or other alarm statuses in the FRTU box to determine if the device is still alive. The analog measurements are embedded in a sensor that is equipped with transducers to measure current and voltage in milliamp or voltage values. To process the values on an FRTU/RTU, conversion to digital values on analog-to-digital (A/D) converter takes place. There are typically four types of binary status: (i) current status or state, (ii) current status with memory detection, which is the number of contact changes since the last report to the master, (iii) sequence-of-events (SOE) logs where the contact changes are tabulated with respect to time of occurrence, and (iv) accumulator value, which is a counter over a period of time [95, 98].

The remote control may also involve using interposing relays to achieve a control action commanded by the master. In the technology today, the intent is to facilitate remote control and setup so that it can be centralized in the management and regular maintenance of all FRTUs in the distribution network. The desire is to follow the control decision closer to the operating point so that the computational process can be reduced on the master servers as well as on the communication system based on the measurements from FRTUs and RTUs. Additional implementation includes closed-loop control, which is achieved within the RTU/FRTU itself by comparing a measured quantity to a setpoint downloaded from the master. If the measured quantity is low, a remedial control action is sent to the device in order to increase the value. Usually, a permissible error is defined to limit excessive control action. Closed-loop control mechanisms have been implemented on load-tap changing transformers, voltage regulators, etc. [98].

The next level of complexity is to use several measured values to compute the desired result, compare the results to the desired value, and then execute appropriate control actions. This procedure might be used to control a number of capacitor banks on a feeder using either voltage measurements along the feeder, reactive flow in the feeder, or the power factor of the feeder current. The further level of complexity extends the computation to optimizing the measured or calculated quantity. The optimization might be to minimize or maximize the given quantity [95, 98].

4.4.2 Incorporation of Metering Datasets from a Secondary Network

As a foundation of the modern grid, an advanced metering infrastructure (AMI) is able to collect the consumption information (electric, gas, or water) from the users' side through remote communications. The smart meter is the basic component in an AMI that measures consumption in more detail and communicates that consumption information to the utility and customer [99–102].

In contrast to monthly consumption meter reads, metering data refers to consumption data measured and recorded by smart meters at more frequent intervals, such as once per hour or every 15 min. The metering data management collects the captured information from each meter associated with the time tags from the secondary network and uses the data as a reference to resolve billing issues. The time-series data is also critical to maintaining, data integrity, since the data is shared across multiple utilities such as demand response (DR), forecasting, distribution-asset analysis, etc. [99, 101].

Since meter reading systems in the secondary network can vary widely in reading frequency and overlap in a geographic area, a user should have the right to define the individual rules and logic of the metering measurements for processing "questionable" reads, where data appears to be inaccurate [101]. The meter data management should enable users to define the proper reading via multi-overlapping readings within the same request window. If a valid reading cannot be obtained within the reading window, the meter data management can derive an acceptable reading according to archived or historical data or resend a new read request using a variety of service rules [103].

4.5   Data Structure for Power Flow

Power flow analysis in the distribution system plays a critical role in automation algorithms of radial topology, which include the proposal of fault isolation, the combination of switch sets in network reconfiguration, and service restoration. A large-scale radial distribution system has a complicated topology and is subject to frequent changes for maintenance, which includes load balancing and emergency operations. Therefore, a dynamic data structure that permits dynamic topology reconfiguration, isolation, and restoration in a fault event is necessary for power flow analysis. The ability of automation in the modern power system to handle these complex operations, which require frequent topology changes (topological reconfigurations) in the distribution system, demands a dynamic topology recorder based on a well-defined data structure [104, 105].

The following attributes were envisioned while developing a data structure for a distribution system:

- The data structure required for holding the information of each component in the system must be compact and addressable from any function [104].

- It must be flexible, to accommodate any number of vertices and edges (branches) in a system.

- It must contain enough information for relevant calculations and power flow analysis.

In a MATLAB programming environment, a data structure can be treated as a structure array,which is a data type that groups related data using data containers called fields. Each field can contain data of any type and size [106]. The **struct** function can create a structure array. Multiple fields can be specified simultaneously to create a nonscalar structure array. The syntax of **struct** for a single field is:

$$\text{structure} = \textbf{struct}(\text{field,value}),$$

and the multiple fields syntax is:

$$\text{structure} = \textbf{struct}(\text{field1,value1,...,fieldN,valueN}).$$

You can access data in a field using dot notation of the form **structName.fieldName**. The detailed examples will be shown later.

4.5.1 Components in a Distribution System

The estimation of power flow at each node in a distribution feeder is the basis for judging the normality of lines and equipment. The voltages should be determined and maintained within specified limits. Because the size and complexity of loads in a system continue to grow, the relevant computations become more and more burdensome. In order to simulate the power flow in the system more accurately, each system component should be defined and simulated in more detail [79, 107] as follows:

- Simulate multi-voltage primary circuits supplying a large secondary network and multiple isolated networks simultaneously in an integrated topology;

- Simulate three-phase transformers with various connection types;

- Simulate composite loads, including the voltage-dependent effects;

- Analyze the system loss and detailed losses for each branch;

- Analyze emergency cases;

- Analyze single-, two-, and three-phase unbalanced systems simultaneously;

- Analyze the effects of voltage regulations;

- Analyze the effects of shunt capacitors;

- Analyze the effects of voltage regulators.

To achieve these purposes, the corresponding data types and parameters for each component in a distribution system should be defined:

- Load.

  – Connection Types: delta or wye.
  – Parameters: kVA, power factor, constant power, constant current, and constant impedance.

- Capacitor.

  – Connection Types: delta or wye.
  – Parameters: kVAr and kV.

- Regulator.

  – Types: type A or type B.
  – Connection Types: delta or wye.
  – Parameter: tap.

- Branch: Overhead or Underground.

  – Overhead Parameters: phase conductor, neutral conductor, Cartesian coordinates of abcn.
  – Underground: tape shield or concentric neutral.
  – Tape Shield Parameters: cable data, thickness of tape shield, neutral data, and phase conductor diameter.
  – Concentric Neutral Parameters: outside diameter, number of strands, phase conductor, neutral conductor, radius of the circle passing through the center of the strands, Cartesian coordinates of abcn.

- Transformers [108]

  – Types: delta grounded wye, ungrounded wye delta, open wye open delta, ground wye wye, delta delta, and open delta delta.
  – Parameters for Each Type of Transformer: kVLL high, kVLN low, kVA, Z.

Listing 4.1 demonstrates examples of data structure for components in a distribution system. The components include load, capacitor, regulator, transformers, and overhead and underground branches.

The loads defined here are three-phase unbalanced with delta or wye connection type. The regulator has two types: "A" and "B." The phase conductor of the overhead line is selected as *336,400 26/7 ACSR*, while the neutral conductor is *4/0 6/1 ACSR*. To define the parameters for the series impedance of the concentric neutral, the number of strands is set as 13, the phase conductor is selected as the type *250,000 AA*, while the neutral conductor is *14 copper*. In the setting of tape shield, the cable data is chosen from *1/0 AA* and the neutral data is selected as *1/0 copper, 7 strands*. The types of transformers include the "delta grounded wye," "undergrounded wye delta," "open wye open delta," "grounded wye wye," "delta delta," and "open delta delta."

Listing 4.1
Example of Data Structure for Components in a Distribution System.

```
 1   %%======================= LOAD =============================
 2   cload31 = 'kVA';                      clv31 = {900, 1250, 1000};
 3   cload32 = 'power_factor';             clv32 = {0.9, 0.95, 0.85};
 4   load3 = struct(cload31,clv31,cload32,clv32);
 5
 6   cload51 = 'kVA';                      clv51 = {1100, 1200, 1300};
 7   cload52 = 'power_factor';             clv52 = {0.9, 0.9, 0.9};
 8   load5 = struct(cload51,clv51,cload52,clv52);
 9
10   cload61 = 'kVA';                      clv61 = {1000, 1200, 1000};
11   cload62 = 'power_factor';             clv62 = {0.9, 0.95, 0.9};
12   load6 = struct(cload61,clv61,cload62,clv62);
13
14   tload1 = 'load3';       tlv1 = load3;
15   tload2 = 'load5';       tlv2 = load5;
16   tload3 = 'load6';       tlv3 = load6;
17   tload = struct(tload1,tlv1,tload2,tlv2,tload3,tlv3);
18
19   %Types
20   ctload1 = 'Δ';    ctlv1 = tload;
21   ctload2 = 'wye';       ctlv2 = tload;
22   ctload = struct(ctload1,ctlv1,ctload2,ctlv2);
23
24   %Define Load Parameters
25   load = struct('connection',ctload);
26
27   %%=================== Capacitor =============================
28   %Three Phases
29   ctp1 = 'kVAr';                      ctplv1 = {600, 600, 600};
30   ctp2 = 'kV';                        ctplv2 = {12.47};
31   ctp = struct(ctp1,ctplv1,ctp2,ctplv2);
32
33   %Types
34   ctload1 = 'Δ';    ctlv1 = ctp;
35   ctload2 = 'wye';       ctlv2 = ctp;
36   ctload = struct(ctload1,ctlv1,ctload2,ctlv2);
37
38   %Define Load Parameters
39   capacitor = struct('connection',ctload);
40
41   %%=================== Regulator =============================
42   %Three Phases
43   tr1 = 'tap'; trv1 = {9, 7, 2};
44   tr = struct(tr1,trv1);
45
46   %Connection
47   rc1 = 'Δ'; rcv1 = tr;
48   rc2 = 'wye';   rcv2 = tr;
49   rc = struct(rc1,rcv1,rc2,rcv2);
50
51   %Types
52   ret1 = 'typeA'; retv1 = rc;
53   ret2 = 'typeB'; retv2 = rc;
54   ret = struct(ret1,retv1,ret2,retv2);
```

```
55    regulator = struct('types',ret);
56
57
58    %%=========================== Branch  ===============================
59
60    % Series Impedance of Overhead Line
61    % phase conductor:336,400 26/7 ACSR
62    ohl1 = 'GMRi';    ohlv1 = {0.0244};
63    ohl2 = 'ri';      ohlv2 = {0.306};
64    ohl3 = 'RDc';     ohlv3 = {0.03004};
65    % neutral conductor:4/0 6/1 ACSR
66    ohl4 = 'GMRn';    ohlv4 = {0.00814};
67    ohl5 = 'rn';      ohlv5 = {0.592};
68    ohl6 = 'RDn';     ohlv6 = {0.02346};
69    % Cartesian coordinates of abcn
70    ohl7 = 'd_a';     ohlv7 = {0+29i};
71    ohl8 = 'd_b';     ohlv8 = {2.5+29i};
72    ohl9 = 'd_c';     ohlv9 = {7+29i};
73    ohl10 = 'd_n';    ohlv10 = {4+25i};
74    ohl11 = 'w';      ohlv11 = {376.9911};
75    ohl = struct(ohl1,ohlv1,ohl2,ohlv2,ohl3,ohlv3,...
76          ohl4,ohlv4,ohl5,ohlv5,ohl6,ohlv6,ohl7,ohlv7,...
77          ohl8,ohlv8,ohl9,ohlv9,ohl10,ohlv10);
78
79    % Series Impedance of Concentric Neutral
80    % Outside diameter
81    ugl1 = 'd_od';      uglv1 = {1.29};
82    % 13 strands
83    ugl2 = 'k';         uglv2 = {13};
84    % Phase Conductor:250,000 AA
85    ugl3 = 'GMRp';      uglv3 = {0.0171};
86    ugl4 = 'dp';        uglv4 = {0.567};
87    ugl5 = 'rp';        uglv5 = {0.41};
88    % Neutral Conductor:#14 copper
89    ugl6 = 'GMRs';      uglv6 = {0.00208};
90    ugl7 = 'rs';        uglv7 = {14.87};
91    ugl8 = 'ds';        uglv8 = {0.0641};
92    % Radius of the circle passing through the center of the strands
93    ugl15 = 'R';        uglv15 = {0.0511};
94    % Cartesian coordinates of three phases
95    ugl9 = 'd1';        uglv9 = {0, 0, 0};
96    ugl10 = 'd2';       uglv10 = {0, 0.5, 0.5};
97    ugl11 = 'd3';       uglv11 = {0, 0, 1+1i};
98    ugl12 = 'd4';       uglv12 = {0, 0.0511i, 0.0511i};
99    ugl13 = 'd5';       uglv13 = {0, 0.5+0.0511i, 0.5+0.0511i};
100   ugl14 = 'd6';       uglv14 = {0, 0, 1+0.0511i};
101   ugl =struct(ugl1,uglv1,ugl2,uglv2,ugl3,uglv3,...
102         ugl4,uglv4,ugl5,uglv5,ugl6,uglv6,ugl7,uglv7,...
103         ugl8,uglv8,ugl15,uglv15,ugl9,uglv9,ugl10,...
104         uglv10,ugl11,uglv11,ugl12,uglv12,ugl13,uglv13,ugl14,uglv14);
105
106   %Tape Shield
107   % cable data:1/0 AA
108   uts1 = 'ds';        utsv1 = {0.88};
109   uts2 = 'r_cable';   utsv2 = {0.97};
110   uts3 = 'GMRp';      utsv3 = {0.0111};
111   % thickness of tape shield
```

```
112  uts4 = 'T';              utsv4 = {0.005};
113  % neutral data:1/0 copper, 7 strand
114  uts5 = 'r_neutral'; utsv5 = {0.607};
115  uts6 = 'GMRn';            utsv6 = {0.01113};
116  uts7 = 'Dnm';            utsv7 = {3};
117  % phase conductor diameter
118  uts8 = 'dp';              utsv8 = {0.368};
119  uts = struct(uts1,utsv1,uts2,utsv2,uts3,utsv3,...
120      uts4,utsv4,uts5,utsv5,uts6,utsv6,uts7,utsv7,uts8,utsv8);
121
122  %Underground Types
123  unty1 = 'concentric_neutral'; untyv1 = ugl;
124  unty2 = 'tape_shield';          untyv2 = uts;
125  unty = struct(unty1,untyv1,unty2,untyv2);
126
127  %Types
128  branty1 = 'overhead';          brantyv1 = ohl;
129  branty2 = 'underground';       brantyv2 = unty;
130  branch = struct(branty1,brantyv1,branty2,brantyv2);
131
132
133  %%====================== Transformers  ======================
134
135  %Delta Grounded Wye
136  dgy1 = 'kVLL_high';          dgyv1 = {12.47};
137  dgy2 = 'kVLN_low';          dgyv2 = {2.4};
138  dgy3 = 'transkVA';          dgyv3 = {5000};
139  dgy4 = 'Z';              dgyv4 = {0.01+0.06i};
140  dgy5 = 'AV';              dgyv5 = {[0 -1 0;0 0 -1;-1 0 0]};
141  dgy6 = 'W';              dgyv6 = {[2 1 0;0 2 1;1 0 2]/3};
142  dgy7 = 'D';              dgyv7 = {[1 -1 0;0 1 -1;-1 0 1]};
143  dgy8 = 'AI';              dgyv8 = {[1 0 0;0 1 0;0 0 1]};
144  dgy = struct(dgy1,dgyv1,dgy2,dgyv2,dgy3,dgyv3,dgy4,dgyv4,...
145      dgy5,dgyv5,dgy6,dgyv6,dgy7,dgyv7,dgy8,dgyv8);
146
147  %Ungrounded Wye Delta
148  uyd1 = 'kVLL_high';          uydv1 = {7200};
149  uyd2 = 'kVLN_low';          uydv2 = {240};
150  uyd3 = 'transkVA';          uydv3 = {100};
151  uyd4 = 'Z';              uydv4 = {1+4i};
152  uyd5 = 'AV';              uydv5 = {[1 0 0;0 1 0;0 0 1]};
153  uyd6 = 'AI';              uydv6 = {[1 0 0;0 1 0;0 0 1]};
154  uyd7 = 'W';              uydv7 = {[2 1 0;0 2 1;1 0 2]/3};
155  uyd8 = 'D';              uydv8 = {[1 -1 0;0 1 -1;-1 0 1]};
156  uyd9 = 'L';              uydv9 = {[1 -1 0;1 2 0;-2 -1 0]/3};
157  uyd = struct(uyd1,uydv1,uyd2,uydv2,uyd3,uydv3,uyd4,uydv4,...
158      uyd5,uydv5,uyd6,uydv6,uyd7,uydv7,uyd8,uydv8,uyd9,uydv9);
159
160  %Open Wye Open Delta
161  oyd1 = 'kVLL_high';          oydv1 = {7200};
162  oyd2 = 'kVLN_low';          oydv2 = {240};
163  oyd3 = 'transkVA';          oydv3 = {100};
164  oyd4 = 'Z';              oydv4 = {1+4i};
165  oyd5 = 'AV';              oydv5 = {[1 0 0;0 1 0;0 0 0]};
166  oyd6 = 'AI';              oydv6 = {[1 0 0;-1 1 0;0 -1 0]};
167  oyd7 = 'W';              oydv7 = {[2 1 0;0 2 1;1 0 2]/3};
168  oyd8 = 'D';              oydv8 = {[1 -1 0;0 1 -1;-1 0 1]};
```

```
169   oyd9 = 'b';                      oydv9 = {[1 0 0;0 0 -1;0 0 0]};
170   oyd10 = 'd';                     oydv10 = {[1 0 0;0 0 -1;0 0 0]};
171   oyd11 = 'BV';                    oydv11 = {[1 0 0;0 1 0;-1 -1 0]};
172   oyd12 = 'z';                     oydv12 = {[1 0 0;0 0 -1;-1 0 1]};
173   oyd = struct(oyd1,oydv1,oyd2,oydv2,oyd3,oydv3,oyd4,oydv4,...
174        oyd5,oydv5,oyd6,oydv6,oyd7,oydv7,oyd8,oydv8,oyd9,...
175        oydv9,oyd10,oydv10,oyd11,oydv11,oyd12,oydv12);
176
177   %Grounded Wye Wye
178   gyy1 = 'kVLL_high';              gyyv1 = {7200};
179   gyy2 = 'kVLN_low';               gyyv2 = {240};
180   gyy3 = 'transkVA';               gyyv3 = {100};
181   gyy4 = 'Z';                      gyyv4 = {1+4i};
182   gyy5 = 'AV';                     gyyv5 = {[1 0 0;0 1 0;0 0 1]};
183   gyy6 = 'AI';                     gyyv6 = {[1 0 0;0 1 0;0 0 1]};
184   gyy = struct(gyy1,gyyv1,gyy2,gyyv2,gyy3,gyyv3,gyy4,...
185        gyyv4,gyy5,gyyv5,gyy6,gyyv6);
186
187   %Delta Delta
188   dd1 = 'kVLL_high';               ddv1 = {12470};
189   dd2 = 'kVLN_low';                ddv2 = {240};
190   dd3 = 'transkVA';                ddv3 = {100};
191   dd4 = 'Z';                       ddv4 = {1+4i};
192   dd5 = 'AV';                      ddv5 = {[1 0 0;0 1 0;0 0 1]};
193   dd6 = 'AI';                      ddv6 = {[1 0 0;0 1 0;0 0 1]};
194   dd7 = 'DI';                      ddv7 = {[1 0 -1;-1 1 0;0 -1 1]};
195   dd8 = 'W';                       ddv8 = {[2 1 0;0 2 1;1 0 2]/3};
196   dd9 = 'D';                       ddv9 = {[1 -1 0;0 1 -1;-1 0 1]};
197   dd = struct(dd1,ddv1,dd2,ddv2,dd3,ddv3,dd4,ddv4,dd5,ddv5,...
198        dd6,ddv6,dd7,ddv7,dd8,ddv8,dd9,ddv9);
199
200   %Open Delta Delta
201   odd1 = 'kVLL_high';              oddv1 = {12470};
202   odd2 = 'kVLN_low';               oddv2 = {240};
203   odd3 = 'transkVA';               oddv3 = {100};
204   odd4 = 'Z';                      oddv4 = {1+4i};
205   odd5 = 'AV';                     oddv5 = {[1 0 0;0 1 0;-1 -1 0]};
206   odd6 = 'AI';                     oddv6 = {[1 0 0;-1 0 -1;0 0 1]};
207   odd7 = 'W';                      oddv7 = {[2 1 0;0 2 1;1 0 2]/3};
208   odd8 = 'D';                      oddv8 = {[1 -1 0;0 1 -1;-1 0 1]};
209   odd = struct(odd1,oddv1,odd2,oddv2,odd3,oddv3,odd4,oddv4,...
210        odd5,oddv5,odd6,oddv6,odd7,oddv7,odd8,oddv8);
211
212   %Types
213   transty1 = 'Delta_Grounded_Wye'; transtyv1 = dgy;
214   transty2 = 'Ungrounded_Wye_Delta'; transtyv2 = uyd;
215   transty3 = 'Open_Wye_Open_Delta'; transtyv3 = oyd;
216   transty4 = 'Grounded_Wye_Wye';   transtyv4 = gyy;
217   transty5 = 'Delta_Delta';        transtyv5 = dd;
218   transty6 = 'Open_Delta_Delta';   transtyv6 = odd;
219   transty = struct(transty1,transtyv1,transty2,transtyv2,...
220        transty3,transtyv3,transty4,transtyv4,transty5,...
221        transtyv5,transty6,transtyv6);
222
223   transformer = struct('types', transty);
```

4.5.2 Generate Relevant Matrices for Power Flow Analysis

The process to store detailed parameters and information of each component into a data structure is to save the storage space and accelerate calculating speed. However, the data cannot be applied to the power flow analysis directly since there still need to be some transition calculations. For example, the representation of a load in the forward and backward sweep should be a load matrix. A conversion needs to extract and combine all relative information, such as the load value for each phase, connection type, and the power factor to generate the matrix form. Listing 4.2 illustrates the detailed transition process of load matrix in a MATLAB environment.

Listing 4.2
Generate Load Matrix

```
1   % Load and line-line & line-neutral voltage
2   a3 = load.connection.wye.load3(1).kVA;
3   b3 = load.connection.wye.load3(2).kVA;
4   c3 = load.connection.wye.load3(3).kVA;
5
6   aa3 = acosd(load.connection.wye.load3(1).power_factor);
7   ab3 = acosd(load.connection.wye.load3(2).power_factor);
8   ac3 = acosd(load.connection.wye.load3(3).power_factor);
9
10  a3 = a3*cos(aa3)+a3*sin(aa3)*1i;
11  b3 = b3*cos(ab3)+b3*sin(ab3)*1i;
12  c3 = c3*cos(ac3)+c3*sin(ac3)*1i;
13
14  S3 = [a3;b3;c3] * 1000;
15
16  a5 = load.connection.wye.load5(1).kVA;
17  b5 = load.connection.wye.load5(2).kVA;
18  c5 = load.connection.wye.load5(3).kVA;
19
20  aa5 = acosd(load.connection.wye.load5(1).power_factor);
21  ab5 = acosd(load.connection.wye.load5(2).power_factor);
22  ac5 = acosd(load.connection.wye.load5(3).power_factor);
23
24  a5 = a5*cos(aa5)+a5*sin(aa5)*1i;
25  b5 = b5*cos(ab5)+b5*sin(ab5)*1i;
26  c5 = c5*cos(ac5)+c5*sin(ac5)*1i;
27
28  S5 = [a5;b5;c5] * 1000;
29
30  a6 = load.connection.wye.load6(1).kVA;
31  b6 = load.connection.wye.load6(2).kVA;
32  c6 = load.connection.wye.load6(3).kVA;
33
34  aa6 = acosd(load.connection.wye.load6(1).power_factor);
35  ab6 = acosd(load.connection.wye.load6(2).power_factor);
36  ac6 = acosd(load.connection.wye.load6(3).power_factor);
37
38  a6 = a6*cos(aa5)+a6*sin(aa5)*1i;
39  b6 = b6*cos(ab5)+b6*sin(ab5)*1i;
```

```
40  c6 = c6*cos(ac5)+c6*sin(ac5)*1i;
41
42  S6 = [a6;b6;c6] * 1000;
43
44  I_value{3} = conj(S3./V_value{3});
45  I_value{5} = conj(S5./V_value{5});
46  I_value{6} = conj(S6./V_value{6});
```

Listing 4.3 shows the process of generating a capacitor matrix with a wye connection. Listing 4.4 is an example that shows the generation of relevant matrices of a "delta grounded wye" transformer.

Listing 4.3
Generate Capacitor Matrix

```
1   ac = capacitor.connection.wye(1).kVAr;
2   bc = capacitor.connection.wye(2).kVAr;
3   cc = capacitor.connection.wye(3).kVAr;
4   dc = capacitor.connection.wye.kV;
5
6   temp1 = ac/(dc^2*1000);
7   temp2 = bc/(dc^2*1000);
8   temp3 = cc/(dc^2*1000);
9   temp4 = [temp1 temp2 temp3]';
10
11  captemp = temp4 * 1i;
12
13  for i = 1:6
14      capstr{i,1} = zeros(3,1);
15  end
16
17  capstr{2} = captemp;
```

Listing 4.4
Generate Transformer Matrix

```
1   % Delta_Grounded_Wye transformer
2   nt = transformer.types.Delta_Grounded_Wye.kVLL_high...
3       /(transformer.types.Delta_Grounded_Wye.kVLN_low);
4   z_base = ((transformer.types.Delta_Grounded_Wye.kVLN_low)^2)...
5       *1000/transformer.types.Delta_Grounded_Wye.transkVA;
6   zt = transformer.types.Delta_Grounded_Wye.Z * z_base;
7   At = 1/nt * [1 0 -1;-1 1 0;0 -1 1];
8   Bt = [zt 0 0;0 zt 0;0 0 zt];
9   dt = 1/nt * transformer.types.Delta_Grounded_Wye.D;
10  at = -nt/3 * [0 2 1;1 0 2;2 1 0];
```

Relatively speaking, the generation of the impedance matrix of a branch is more complex than others. As shown in Listing 4.5, taking an overhead line as an example, to calculate the phase impedance matrix of the segment, the distances between the positions need to be computed. Apply the modified Carson's equation to calculate the self and mutual impedance between phases, and then use the "Kron" reduction based on the primitive impedance-matrix partitioned form to get the result. In addition, the image distances and the self/mutual potential coefficients need to be considered for computing the phase admittance matrix. It should be noted that the elements of the phase admittance matrix are very small [41]. The "a," "b," "c," "d," "A," and "B" matrices can be generated. During the calculation, the distance units (miles, feet, or meters) should be noted.

Listing 4.5
Generate Branch Impedance Matrix

```
1   % Distance
2   D_ab = abs(branch.overhead.d_a - branch.overhead.d_b);
3   D_bc = abs(branch.overhead.d_b - branch.overhead.d_c);
4   D_ca = abs(branch.overhead.d_a - branch.overhead.d_c);
5   D_an = abs(branch.overhead.d_a - branch.overhead.d_n);
6   D_bn = abs(branch.overhead.d_b - branch.overhead.d_n);
7   D_cn = abs(branch.overhead.d_c - branch.overhead.d_n);
8
9   % Impedance
10  z_aa = 0.0953 + branch.overhead.ri + 0.12134i * ...
        (log(1/branch.overhead.GMRi)+7.93402);
11  z_nn = 0.0953 + branch.overhead.rn + 0.12134i * ...
        (log(1/branch.overhead.GMRn)+7.93402);
12  z_ab = 0.0953 + 0.12134i * (log(1/D_ab)+7.93402);
13  z_bc = 0.0953 + 0.12134i * (log(1/D_bc)+7.93402);
14  z_ac = 0.0953 + 0.12134i * (log(1/D_ca)+7.93402);
15  z_an = 0.0953 + 0.12134i * (log(1/D_an)+7.93402);
16  z_bn = 0.0953 + 0.12134i * (log(1/D_bn)+7.93402);
17  z_cn = 0.0953 + 0.12134i * (log(1/D_cn)+7.93402);
18
19  z_ij = [z_aa,z_ab,z_ac;z_ab,z_aa,z_bc;z_ac,z_bc,z_aa];
20  z_in = [z_an;z_bn;z_cn];
21  z_nj = [z_an,z_bn,z_cn];
22
23  z_abc = z_ij - z_in * ((z_nn)^(-1)) *z_nj;
24
25  % Image distance
26  S_aa = abs(branch.overhead.d_a - conj(branch.overhead.d_a));
27  S_bb = abs(branch.overhead.d_b - conj(branch.overhead.d_b));
28  S_cc = abs(branch.overhead.d_c - conj(branch.overhead.d_c));
29  S_ab = abs(branch.overhead.d_a - conj(branch.overhead.d_b));
30  S_bc = abs(branch.overhead.d_b - conj(branch.overhead.d_c));
31  S_ca = abs(branch.overhead.d_c - conj(branch.overhead.d_a));
32  S_an = abs(branch.overhead.d_a - conj(branch.overhead.d_n));
33  S_bn = abs(branch.overhead.d_b - conj(branch.overhead.d_n));
34  S_cn = abs(branch.overhead.d_c - conj(branch.overhead.d_n));
35  S_nn = abs(branch.overhead.d_n - conj(branch.overhead.d_n));
36
37  % Self- and mutual potential coefficients
38  p_aa = 11.17689 * log(S_aa/branch.overhead.RDc);
39  p_bb = 11.17689 * log(S_bb/branch.overhead.RDc);
40  p_cc = 11.17689 * log(S_cc/branch.overhead.RDc);
41  p_ab = 11.17689 * log(S_ab/D_ab);
```

```
42  p_bc = 11.17689 * log(S_bc/D_bc);
43  p_ca = 11.17689 * log(S_ca/D_ca);
44  p_nn = 11.17689 * log(S_nn/branch.overhead.RDn);
45  p_an = 11.17689 * log(S_an/D_an);
46  p_bn = 11.17689 * log(S_bn/D_bn);
47  p_cn = 11.17689 * log(S_cn/D_cn);
48
49  p_ij = [p_aa p_ab p_ca;p_ab p_bb p_bc;p_ca p_bc p_cc];
50  p_in = [p_an;p_bn;p_cn];
51  p_nj = [p_an,p_bn,p_cn];
52
53  p_abc = p_ij - p_in * ((p_nn)^(-1)) *p_nj;
54  y_abc = 376.9911i * (p_abc)^(-1);
55
56  % Source line segment
57  Z_abc = z_abc .* (2000/5280);
58  Y_abc = y_abc .* (2000/5280);
59
60  a1 = [1 0 0;0 1 0;0 0 1]+ (Z_abc.*Y_abc)/2;
61  b1 = Z_abc;
62  c1 = Y_abc + (Y_abc .* Z_abc .* Y_abc)/4;
63  d1 = [1 0 0;0 1 0;0 0 1]+ (Z_abc .* Y_abc)/2;
64  A1 = a1^(-1);
65  B1 = A1 * b1;
66
67
68  % Load line segment
69  ZL_abc = z_abc .* (2500/5280);
70  YL_abc = y_abc .* (2500/5280);
71
72  a2 = [1 0 0;0 1 0;0 0 1]+ (ZL_abc.*YL_abc)/2;
73  b2 = ZL_abc;
74  c2 = YL_abc + (YL_abc .* ZL_abc .* YL_abc)/4;
75  d2 = [1 0 0;0 1 0;0 0 1]+ (ZL_abc .* YL_abc)/2;
76  A2 = a2^(-1);
77  B2 = A2 * b2;
```

4.5.3   Power Flow Programming

The cornerstone of power flow in the radial network is constructing the structure matrix according to the graph representation, which is utilized in carrying out the radial power flow backward/forward iterative steps. The structure matrix is designed to display the connection states, and all of its eigenvalues must be equal to one to ensure its invertibility. By applying this matrix in solving the power flow problem, the data structure with corresponding characteristics and parameters of each node and edge are incorporated.

The distribution power flow technique is flexible in accommodating any changes that may occur in a radial topology, because any topological changes can be exclusively incorporated within the structural matrix. The distribution network power flow is tested by using several balanced and unbalanced three-phase radial distribution systems.

Algorithm 5 is the pseudocode that shows the procedure of the forward and backward sweep power flow analysis based on adjacency matrix and data structure. The adjacency matrix represents the topology of the radial distribution network and the data structure provides the corresponding matrices information. The bus numbering procedure was a sophisticated parent node and child node arrangement. This algorithm requires:

- Adjacency matrix M_a of network topology with optimal ordering.

- Line distance structure.

- Data structure of distribution network components.

- Extract data from structure and calculate generized matrices for each element the "a," "b," "c," "d," "A," and "B" matrices.

- Compute load current.

- Assume node with number index 1 as the source node.

    The algorithm is summarized as follows:

1. Require the topological adjacency matrix M and the source voltage V_s;

2. Perform the BFS on M and save the results with different levels into a new data structure "lvl{}";

3. Initial the voltage V and current I before the iteration as 0s;

4. Define the source voltage and store the value into the first level;

5. Set the criteria and threshold of the sweep;

6. Sweep each level from the second level to the outermost level;

7. Calculate the load current;

8. Find each element's parent index to extract corresponding values from the structure;

9. Do backward and forward sweep;

10. Return the voltage for all nodes V and currents for all nodes I.

Listing 4.6
Matrix-Based Backward and Forward Power Flow

```
1  clc
2  clear all
3  close all
4
5  % Load datastrucsture and calculate a,b,c,d,A,B
6  % shortc_ABd
```

Algorithm 5 Graph-Based Power Flow

Input:
   $\mathbf{M_a}, \mathbf{V_s}$, "c", "d", "A", and "B"Matrices
Output:
   Voltage for all nodes $\mathbf{V}$, Currents for all nodes $\mathbf{I}$
1: % Initial voltage of feeder tail;
2: $V_{old} \leftarrow 0$;
3: **for $\mathbf{M_a}$ do**
4:    % Perform breadth-first search (BFS);
5:    $[\mathbf{C}, \mathbf{Indx}] = \mathbf{sort(bfsadj(M_a, 1))}$;
6:    % Store index of vertices within a same level into a structure;
7:    $\mathbf{X = unique(C)}$;
8:    **for i = 1 : length(X) do**
9:      $\mathbf{tempIndx = find(C == X(i))}$;
10:     $\mathbf{lvl\{i, 1\} \leftarrow Indx(tempIndx)}$;
11:    **end for**
12:   % Initialize two empty data structures $\mathbf{V, I}$;
13:   $\mathbf{V} \leftarrow 0$;
14:   $\mathbf{I} \leftarrow 0$;
15:   % Define source voltage;
16:   $\mathbf{lvl\{1\} \leftarrow V_s}$;
17:   % Set the criteria and initialize iteration limits;
18:   **while $((\mathbf{V_{new} - V_{old})/V_{nom} > 0.005})\&(l < 1000)$ do**
19:    $i \leftarrow \mathbf{2 : length(lvl)}$;
20:    $j \leftarrow \mathbf{lvl\{i\}}$;
21:    **for n = 1 : size(lvl{i}) do**
22:      $m = \mathbf{j(n)}$;
23:      % Compute load current I_L;
24:      $\mathbf{I\{load\} \leftarrow I_L}$;
25:      % Find each vertex' parent and children node(s);
26:      **if Number of children nodes $\mathbf{N_c} = 1$ then**
27:        $\mathbf{I\{pare(m)\} = c * V\{m\} - d * I\{m\}}$;
28:      **else**
29:        $\mathbf{I\{pare(m)\} = c * V\{m\} - d * (\sum_{m=1}^{N_c} I\{m\})}$;
30:      **end if**
31:      $\mathbf{V\{m\} = A * V\{parent(m)\} - B * I\{m\}}$;
32:    **end for**
33:   **end while**
34: **end for**
35: **return $\mathbf{V, I}$**

---

7 ABcd
8 testdata
9

```
10    % Initial value of feeder tail
11    V_old = [0;0;0];
12
13    % BFS to get depth and index of each node
14    [C,I] = sort(bfsadj(AdjM,1));
15
16    % Store the index of nodes in same level in structure
17    X=unique(C);
18    for i=1:length(X)
19        tempI = find(C==X(i));
20        lvl{i,1} = I(tempI);
21    end
22
23    % Initial structure of V and I
24    for i = 1:length(C)
25        V_value{i,1} = zeros(3,1);
26        I_value{i,1} = zeros(3,1);
27    end
28
29    % Source voltage: Assume the infinite bus voltages are ...
           balanced three phase
30    % of 12.47 kV line-to-line
31    V_value{1} = [7199.6;7199.6*cos(deg2rad(-120))+...
32        7199.6*sin(deg2rad(-120))*1i;...
33        7199.6*cos(deg2rad(120))+7199.6*sin(deg2rad(120))*1i];
34    x1=size(lvl,1);  % how many levels
35    x2=max(lvl{x1});  % max index in the outermost level
36
37    % Initial iterations number:if more than 1000 iterations, ...
           break it
38    for l = 1:1000
39
40    % Forward sweep
41    for i = 2:length(lvl)  % from the 2nd to the outermost level
42        j = lvl{i};  % extract the structure of each level except ...
               the source
43            for n = 1:numel(lvl{i})
44                m = j(n);  % extract each elements in the structure ...
                   of each level
45                pare = parents(AdjM,m);  % find each elements' ...
                   parent index
46                V_value{m} = A{m}*V_value{pare}-B{m}*I_value{m};
47            end
48    end
49
50    % Criteria
51    err = abs(abs(V_value{x2})-abs(V_old))/(7.2*1000);
52
53    if err > 0.05
54
55    V_old = V_value{x2};
56    load_matrix
57
58    % Backward sweep
59    for i = length(lvl):-1:2
60        j = lvl{i};
61        k = numel(lvl{i});
```

```
62    if k == 1
63        pare = parents(AdjM,j);
64        I_value{pare} = c1*V_value{j}+d{j}*I_value{j};
65    else
66        for n = 1:k
67            m = j(n);
68            pare = parents(AdjM,m);
69            child = children(AdjM,pare);
70            child_n = numel(child);
71            I_temp = zeros(3,1);
72            for cn = 1:child_n
73                I_temp = I_temp + ...
                        (c1*V_value{m}+d{m}*I_value{child(cn)});
74            end
75            I_value{pare} = I_temp;
76        end
77    end
78 end
79 else
80      break
81  end
82 end
```

4.6   Flipping between Normally Closed (NC) & Normally Open (NO) Switches

Topological reconfiguration has been the important research subject of optimization for interdependent feeders. Under normal conditions, the configuration of the radial network in an automated distribution system is changed from time to time in order to minimize the load and line losses. During a fault, a modern distribution system with automatic monitoring and control functions can achieve real-time fault identification and isolation. The sectionalizing switches and tie switches can be controlled through pilot-wires. The feeder control system could only detect and isolate faults. Dispatchers could isolate a fault remotely following a fault identification process [109]. Therefore, feeder reconfiguration is an important and usable operation to reduce feeder power losses and improve the security and reliabilty of the whole system. There are a number of normally closed (NC) and normally opened (NO) switches in a distribution system. By changing the open/close status of the tie and sectionalizing switches of the network, load currents can be transferred from feeder to feeder.

The normal state of distribution operation varies from time to time. Therefore, it is critical to evaluate the balance of each feeder with a reasonable amount of loads, which has a mixture of commercial, industrial, or residential combinations. The variation among them can be very dissimilar. Determination of peak loads on substation transformers, individual feeders, or the subsystem of a feeder can help investigation of the load studies for future planning (referred to as non-coincidence of peak). Balancing each feeder with different states of NO switches is essential, and feeder reconfiguration allows us to transfer some loads to other feeders with a subtle change, i.e., move the NO switches to other NC switches. The transfer of loads can be effective if careful studies of the state are done over time. The change of switching status from open to closed or vice versa, while maintaining the radially energized states for all feeders, is optimal from an operational standpoint; this can minimize the number of violations of voltage profiles [110–112] as well total power losses [113–115] in the system.

In a practical distribution network, there are multiple switching options for topological reconfiguration. While making a switching decision, it is sufficient to know the variations in the load and line losses from the previous state to the current state. The exact values of line and load losses are not important [116, 117].

The reconfiguration strategy is based on the assumption that the distribution system is kept in radial topology and that each load is served through only one source. All associated loads under a feeder are served from the substation power supply where it is the point of coupling (slack bus/swing bus/reference bus). Each feeder could have hundreds of loads and might have tens of disconnectors (can be remote-controlled switches, too). If the feeder is connected to multiple other feeders from other substations, the state of normally open (NO) can be changed to normally closed (NC) switches, depending on the loading condition at the time. An example is illustrated in Fig. 4.7 where the NO switches (e_6, e_{12}, and e_{18}) serve as the boundary switches with other feeders. The transfer of loads between these feeders can be switched so long as it maintains its radial configuration. Under the circumstance where an isolation of potential fault is necessary, switches (e_3 and e_{16}) are the NC switches for a feeder that can be opened for that purpose. The possibility includes maintaining the radial structure as well as consideration of the operational constraints. These switches can be upgraded with the remote-controlled capability and features so that the operators from the control center can control and decide the optimality of the switching status. The other example would be for the load at node 11, which can be transferred to Feeder 1 by closing the NO switch e_6 and then opening the sectionalizing switch e_7. Similarly, the loads at nodes 10, 11, and 12 can be switched to Feeder 1 just by closing the tie switch e_6 and then opening the sectionalizing switch e_9.

Under emergency conditions, the urgent need for topological reconfiguration is to isolate faulty segments as well as temporary restoration, as most feeder heads of the coupling point are supported by dual substation transformers (ST) in which the switchgear alleviates the loading condition in case one fails, and the other one picks up their loads. The ST has its nominal value rated in MVA. This is often a temporary arrangement in case fault occurs at the substation. The total connected kVA load for each feeder should sum up as no greater than half of the nominal capacity of ST. The transfer of load can be

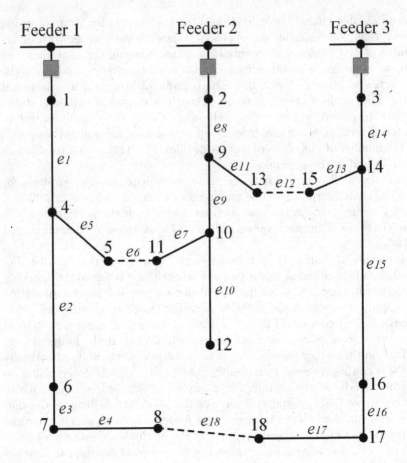

FIGURE 4.7
Three-feeder example for reconfiguration [114].

flexible between the two STs. Similarly, the transfer of loads among all feeders can be reconfigured if there are interdependent NO switches downstream for the distribution substation. The objective is to minimize interruption during abnormal conditions as a result of improving the overall reliability of the electrical distribution grid. This reconfiguration is subject to the operational constraints. Each change of switching of constraints and each fault situation would allow power flow to re-assess the grid and provide all options from the NO switches where operators can make a decision [118].

Once a fault occurs in an automatic distribution system, the faulted area with the faulty component(s) can be isolated by opening the relative remote-controlled switches (RCS). The loads in the isolated area suffer power outage. Under this condition, the distribution network can be divided into three categories: 1) the source (power injection) nodes; 2) unaffected loads with a power supply; 3) the affected loads in the outage area without power sources. The best solution for the service restoration should be created for the loads in the last group. In addition, the following requirements should be enforced during the restoration process:

- minimum switching,

- minimum losses,

- satisfying the voltage and current constraints,

- using switches as close to tie switches as possible.

Without a fault, the amount of loss change resulting from transferring a group of loads from Feeder 1 to Feeder 2 to avoid overload can be estimated from the following simple equation [114, 119, 120]:

$$\Delta P = \text{Re}\left\{ 2\left(\sum_{i\in D} I_i \right)(E_m - E_n)^* \right\} + R_{\text{loop}}| \sum_{i\in D} I_i)|^2$$

where

D is the set of nodes that are disconnected from Feeder 2 and connected to Feeder 1,

m is the tie-switch-connected node of Feeder 1 to which loads from Feeder 2 will be connected,

n is the tie-switch-connected node of Feeder 2 that will be connected to node m via a tie switch,

I_i is the complex current at node i,

R_{loop} is the series resistance of the path connecting the two substation buses of Feeder 1 and Feeder 2 via closure of the specified tie switch,

E_m is the component of $E = R_{\text{node}}I_{\text{node}}$ corresponding to node m: R_{node} is the "node resistance matrix" of Feeder 1 before the load transfer, which is found using the substation bus as reference. I_{node} is the vector of bus currents for Feeder 1.

E_n is similar to E_m, but defined for node n of Feeder 2,

Re. is the function to obtain the real part,

* complex conjugate,

|.| magnitude operators.

Suppose the load at node 11 is transferred from Feeder 2 to Feeder 1 by closing the tie switch e_6 and opening the sectionalizing switch e_7. In this case, $D = 11$, $m = 5$, $n = 11$, and

$$\Delta P = \text{Re}\left\{ 2I_{11}(E_5 - E_{11})^* \right\} + R_{loop}|I_{11}|^2$$

where R_{loop} is the total resistance of the path along the branches e_1, e_5, e_6, e_7, e_9, and e_8.

Suppose the loads at nodes 10, 11, and 12 are transferred from Feeder 2 to Feeder 1 by closing tie switch e_6 and opening the sectionalizing switch e_9. In this case, $D = 10, 11, 12$, $m = 5$, $n = 11$, and

$$\Delta P = \text{Re}\left\{ 2(I_{10} + I_{11} + I_{12})(E_5 - E_{11})^* \right\} + R_{\text{loop}}|I_{10} + I_{11} + I_{12}|^2$$

where R_{loop} is the same as above.

4.7 Conclusions

The unbalanced power flow module is the essential part of the application. Upon a converged solution, it determines the state variables, and the information will be useful to predict the voltage profile of a feeder as well as to estimate power flow at a given time. This chapter establishes the MATLAB formulation on the data structure for the sweeping (backward and forward) technique of unbalanced power flow. The detailed modeling of each element can be explored in W. H. Kersting's text. This is the important chapter where it is modeled based on the topology of a distribution network that is based on Chapters 2 and 3. It is a pivotal chapter because the "what-if" solutions of other feasibility of solutions will be provided with total losses and voltage violation possibilities to system dispatchers for any outage decisions to be made quickly. This chapter provides examples in MATLAB, where it is used to identify the substation or pole-mounted device measurements to initiate the power flow module for a solution, as the availability of tie switches to the neighboring feeders would provide options. Each option is verified by the power-flow analysis, to determine the total losses hypothetically, as well the number of voltage violations. This is to help dispatchers in the control center to identify the best option for partial service restoration while the crew is at the site repairing the circuit. The next chapter will provide identification of fault segments based on the fault indicators of FRTUs and RTUs.

Mini Project 3: Conversion from GIS Datasets to Data Structure

This chapter provides the data structure in MATLAB format where it is used to perform iterative backward-forward sweeping power unbalanced flow. Utilitize the examples of MATLAB script in this chapter to establish the connection. Make sure the script is written in a function with the topology as the input and output with convergent/diverged solutions. Some details of distribution system modeling can refer to W. H. Kersting's textbook on distribution system analysis.

5

Fault Identification Based on Segment Localization

Fault localization is a tricky subject, as it all depends on the availability of SCADA binary status. This highly relies on the vendor products to capture the fault, and how to conclude is based on the sequence of events that is captured by the FRTU/RTU. This chapter introduces the notion of capturing the fault segments based on fault indicators that are part of the remote-controlled switches (RCSs). These RCSs can be operated by dispatchers in the control center to set the values of open/closed remotely. First, the topology of the feeders must be up to date. This is crucial, as the latest status of switches telemetered back from the FRTU will provide the topological update to the database that can be inferred and concluded. Most likely, the relays associated with reclosers or breakers (also RCSs) might de-energize large areas and cause massive outages. At the same times, alarm messages are created where it can be overwhelming to the dispatcher to pre-screen information and understand the scenarios. The sequence of an event may mislead the operator if the message is delayed or lost due to communication problems, and the faulted area may not be accurately detected and conclusive. Therefore, handling different types of faults offers a great deal of possibilities in automation [121–123].

The main tasks of accurate fault analysis to assist operators in these situations include fault types, permanent or temporary, fault localization, fault isolation, and system restoration. The system operators need to understand the topology well about the controlled network area to perform these tasks efficiently. However, the topology usually consists of lots of components and various connection choices. It is hard to find out the exact fault location accurately and rapidly by limited messages. Without the assistance from system automation, operators and managers have to cooperate with the training simulator in order to be proficient and effective using these functions [124].

The management system of outage due to the fault is one of the major components of the DMS. The traditional protection and management techniques use the phasor components of voltage and current to trip circuit breakers (CBs) or control and operate other types of switching devices. Table 5.1 illustrates the comparison between different communication devices in DMS. The vertical elements in this table show the status of relay existing and sensing capability, and also display the capability of gathering data from a phasor measurement unit (PMU) or FRTU/RTU.

TABLE 5.1
Communication in DMS

	Relay	Sensing Capability	PMU	FRTU/RTU
Circuit Breaker	Yes	Yes	Might	Yes
Recloser	Yes	Yes	Might	Yes
Automated Switch	No	Yes	Might	Yes
Disconnector	No	No	No	No
Fuses	Yes/No	No	No	No

The fault analysis in DMS usually includes:

1. Calculation of a possible electrical fault at a given location of a feeder.

2. Sensitivity studies may be required for the planning engineer to identify the best threshold value for relay settings.

   • Balance between not too sensitive and able to find, if any.

   • Balance between the number of reclosers to be deployed along a feeder to avoid complete feeder outage.

3. User defines a presumed fault location.

4. Relay coordination studies.

5. Switching devices that have relay tripping capability include:

   • Circuit breakers at the substation.

   • Reclosers down the feeder to effectively trip a fault without interrupting a customer. NOT ALL automated switches are armed with relays.

   • Fuses might melt and electrically disconnect the circuit from a feeder/lateral.

An efficient fault localization approach could significantly reduce the outage duration and costs.

5.1   Reserve Engineering — Short Circuit Analysis

Under single or multiple fault conditions, the power system design and analysis studies usually perform short circuit analysis to detect current flows and bus voltages in each section of the whole network. Most short circuit calculations of different types are presumed at an exact location with minor modification of topology. In a practically sized distribution network, this modification of topology can treat a load as the "different" load in the system that can be evaluated using the power flow module. The matrix formation and analysis of impedance [125,126] and admittance [127,128] are described on a deterministic basis. This conventional short circuit analysis is needed to develop information directly related to the local fault conditions; for example, for the design and coordination of SCADA, circuit breakers, or other communication switching components in the distribution system [129].

The short circuit analysis for different phases (single-, double-, and three-phase) can be applied to the balanced or unbalanced system [130]. Like many short circuit/fault analyses, a pre-fault state is first determined. Whether it is a balanced or unbalanced system, the pre-fault state provides the initial condition of the network that can be transformed into a sequence domain for analysis. The symmetrical component method is a way to determine the maximum current hypothetically if a fault occurs in a specific location. This method, however, may not be well suited to a distribution feeder that is inherently unbalanced [131]. A distribution short circuit analysis is integrated with a three-phase power flow analysis to solve various types of single or simultaneous faults, including three-phase faults, three-phase-to-ground faults, line-to-line faults, line-to-line-ground faults, and line-to-ground faults [131,132].

A single event fault is assumed at a location that includes the following types:

Case 1: Three-phase faults (three phases)

Case 2: Three-phase-to-ground faults (three phases)

Case 3: Line-to-line faults (double phases)

Case 4: Line-to-ground fault (single phase)

Resistor value is required for any line-to-ground faults. For cases 1 and 3, the value would be infinity.

In order to simulate an approximate real operating environment and generate the reasonable solutions from the short circuit analysis, some conditions need to be considered:

- Define the mathematical formulation for each component in the system, which include the distributed generators, transformers, feeder lines, shunt capacitors, voltage regulators, and loads.

- Select a suitable numerical method.

- Since the load currents may not be small compared to the short circuit currents, the load currents must be considered in the analysis.

- The load voltage characteristics must be considered in the design of the load model.

Considering the large-scale topology of the system simulation, convergence requirements, and execution time, the analysis must also include the consideration of interrelationships between these conditions, and the impact of the chosen technique [133].

The traditional method to do the short circuit analysis can be summarized as two main steps:

1. Pre-fault condition ← Power flow solutions.

2. Modify the topology and then estimate the fault currents.

In addition, another analysis approach is listed as:

1. Presume the power flow converged for the pre-fault condition.

2. Modify the topology.

3. Validate that new topology configuration with the power flow.

4. Power flow may not converge.

The backward and forward sweep approach can be utilized to calculate the pre- and post-fault bus voltages and current flows. As mentioned before, the upstream current can be obtained by summing up all the downstream currents in the backward sweep. As shown in Fig. 5.1, the current formulations can be represented as:

$$I_1 = I_3 + I_4 + I_5 + I_c,$$

and also

$$I_1 = I_{\text{load1}} + I_{\text{load2}} + I_{\text{load3}} + I_c.$$

Bus voltages can be calculated from the first layer (feeder head) and then moved toward the last layer (feeder tail) in the forward sweep. In Fig. 5.1, the voltage calculation sequence is from the source side (node 1) to the load side (nodes 3, 5, and 6). Detailed transformer models for accurate short circuit analysis results should include [134, 135]:

- representation of various transformer types;

- detailed three-phase representation;

- complete representation of grounded types, which includes neutral grounded, ungrounded, and solid grounded;

- lagging or leading with phase shifting in delta connection;

- open-delta connections;

- two- and three-winding transformers;

- three-phase transformer bank, which includes the single-phase transformer or three-limb core transformer;

- core and copper losses excitation current.

Note that different connection transformers can generate different transformer admittance matrices. The admittance matrix is singular for a delta connection transformer, since the reference node in the delta connection winding is floating [134].

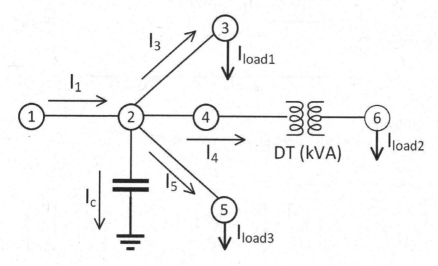

FIGURE 5.1
An example of a radial system.

5.1.1 General Theory

Fig. 5.2 demonstrates a radial system as a model for short circuit analysis. The block represents a three-phase line segment, denoted as Z_{seg}, which is the total equivalent impedance where the faulted bus and equivalent generator are located. The equivalent circuit is the view from the fault node back to the substation. The substation transformer impedance is the external system equivalent impedance combined with the high voltage level of a substation.

Each section of details from the feeder-head down to the rest of the distribution network can be constructed based on the GIS dataset that eventually converts into the incidence matrix that forms the system graph. The equivalent impedance sum of the substation transformer and the external system that is not part of the distribution GIS is provided separately.

Fig. 5.2 also illustrates the impedances, Z_f, that represents the exact fault locations. This value is provided by the planning engineers, with typical fault resistance of 40 ohms selected. This value is a standard value used by many utilities. The voltages E_a, E_b, and E_c, are the equivalent system voltages with presumed 1.0 per unit. The following matrix representation describes

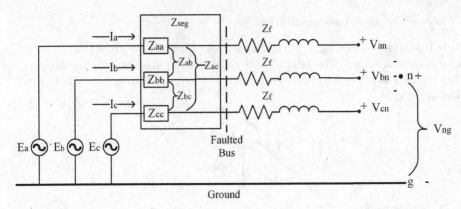

FIGURE 5.2
Radial system model with three-phase line section [131].

the three-phase relationship:

$$\begin{bmatrix} E_a \\ E_b \\ E_c \end{bmatrix} = \begin{bmatrix} Z_{aa} & Z_{ab} & Z_{ac} \\ Z_{ba} & Z_{bb} & Z_{bc} \\ Z_{ca} & Z_{cb} & Z_{cc} \end{bmatrix} \begin{bmatrix} I_a \\ I_b \\ I_c \end{bmatrix} + \begin{bmatrix} Z_f & 0 & 0 \\ 0 & Z_f & 0 \\ 0 & 0 & Z_f \end{bmatrix} \begin{bmatrix} I_a \\ I_b \\ I_c \end{bmatrix} + \begin{bmatrix} I_{an} \\ I_{bn} \\ I_{cn} \end{bmatrix} + \begin{bmatrix} V_{an} \\ V_{bn} \\ V_{cn} \end{bmatrix} + \begin{bmatrix} V_{ng} \\ V_{ng} \\ V_{ng} \end{bmatrix}.$$

Then this equation can be written in compact form by:

$$[E]_{abc} = [Z_{seg}][I]_{abc} + [Z_f][I]_{abc} + [V]_{abcn} + [V]_{ng}$$

and combining terms:

$$[E]_{abc} = \{[Z_{seg}] + [Z_f]\}[I]_{abc} + [V]_{abcn} + [V]_{ng}.$$

Define:

$$[Z]_{com} = [Z_{seg}] + [Z_f].$$

Rearrange terms:

$$[Z]_{com}[I]_{abc} = [E]_{abc} - [V]_{abcn} - [V]_{ng}.$$

Let:

$$[Y_{com}] = [Z_{com}]^{-1}.$$

Then:

$$[I]_{abc} = [Y_{com}][E]_{abc} - [Y_{com}][V]_{abcn} - [Y_{com}][V]_{ng}.$$

Let:

$$[IP]_{abc} = [Y]_{com}[E]_{abc}.$$

Then:

$$[I]_{abc} = [IP]_{abc} - [Y_{com}][V]_{abcn} - [Y_{com}][V]_{ng}.$$

The expended form is:

$$
\begin{bmatrix} I_a \\ I_b \\ I_c \end{bmatrix} = \begin{bmatrix} IP_a \\ IP_b \\ IP_c \end{bmatrix} - \begin{bmatrix} Y_{aa} & Y_{ab} & Y_{ac} \\ Y_{ba} & Y_{bb} & Y_{bc} \\ Y_{ca} & Y_{cb} & Y_{cc} \end{bmatrix} \begin{bmatrix} V_{an} \\ V_{bn} \\ V_{cn} \end{bmatrix} - \begin{bmatrix} Y_{aa} & Y_{ab} & Y_{ac} \\ Y_{ba} & Y_{bb} & Y_{bc} \\ Y_{ca} & Y_{cb} & Y_{cc} \end{bmatrix} \begin{bmatrix} V_{ng} \\ V_{ng} \\ V_{ng} \end{bmatrix}.
$$

Let:

$$Y_a = Y_{aa} + Y_{ab} + Y_{ac}$$

$$Y_b = Y_{ba} + Y_{bb} + Y_{bc}$$

$$Y_a = Y_{ca} + Y_{cb} + Y_{cc}.$$

Then the equations can be rephrased as:

$$I_a = IP_a - (Y_{aa}V_{an} + Y_{ab}V_{bn} + Y_{ac}V_{cn}) - Y_aV_{ng}$$

$$I_b = IP_b - (Y_{ba}V_{bn} + Y_{bb}V_{bn} + Y_{bc}V_{cn}) - Y_bV_{ng}$$

$$I_a = IP_c - (Y_{ca}V_{cn} + Y_{cb}V_{bn} + Y_{cc}V_{cn}) - Y_cV_{ng}.$$

These three equations are the general equations that can be utilized to analyze all types of short circuits. There are seven unknowns ($I_a, I_b, I_c, V_{an}, V_{bn}, V_{cn}$, and V_{ng}). The other three variables (IP_a, IP_b, IP_c) are known functions of the total impedance and system voltages. Four of the seven unknowns must be specified to solve these three equations.

5.1.2 Short Circuit Analysis Methods

The electrical fault usually represents the short circuit in a system. The short circuit analysis can determine the maximum currents under the fault condition. Since the solution of this study can be applied for planning and operating power systems, various types of techniques for short circuit analysis have been developed, as mentioned. The voltage of a bus and the setting of relevant currents where a fault occurs is often preassumed as an initial value for simplicity. However, the bus voltage could be decreased or increased under different load conditions. Strictly speaking, the bus voltage before incurring a fault is not unity. Therefore, this section applies the currents' equations generated from the previous section to analyze the change of currents and voltages when a fault occurred.

5.1.2.1 Three-Phase Fault

The three-phase fault is the symmetrical fault. If the symmetrical fault occurs, the system remains balanced but results in severe damage to the electrical power system equipment. For the line-to-line three-phase fault, the characteristic of the three-phase fault is:

$$V_{an} = V_{bn} = V_{cn} = 0$$

and
$$I_a = I_b = I_c = 0.$$

Then:
$$I_a = IP_a - Y_a V_{ng}$$
$$I_b = IP_b - Y_b V_{ng}$$
$$I_a = IP_c - Y_c V_{ng}.$$

Add these three equations together to solve V_{ng}:

$$V_{ng} = \frac{(IP_a + IP_b + IP_c)}{(Y_a + Y_b + Y_c)}.$$

Therefore, the four necessary unknowns can be used to solve for the three unknown line currents.

For the three-phase line-to-ground fault, the four necessary unknowns can be defined as:
$$V_{an} = V_{bn} = V_{cn} = V_{ng} = 0.$$

5.1.2.2 Two-Phase Fault

The two-phase fault is the unsymmetrical fault, which can also be called the unbalanced fault since it will cause unbalance in the system. The impedance values in each phase that will cause the unbalance currents to flow in different phases are different. The calculation and analysis in an unsymmetrical system are based on a per-phase basis.

For the two-phase line-to-line fault (assume a b-c fault), the characteristic under this condition is:
$$V_{bn} = V_{cn} = 0$$
$$I_a = 0$$
$$I_b + I_c = 0.$$

Using this information, the currents' equations can be rewritten as:

$$I_a = IP_a - Y_{aa} V_{an} - Y_a V_{ng} = 0$$
$$I_b = IP_b - Y_{ba} V_{bn} - Y_b V_{ng}$$
$$I_c = IP_c - Y_{ca} V_{cn} - Y_c V_{ng}.$$

The last two equations are added together, and then rearranging terms yields two equations in matrix form:

$$\begin{bmatrix} IP_a \\ IP_b + IP_c \end{bmatrix} = \begin{bmatrix} Y_{aa} & Y_a \\ (Y_{ba} + Y_{ca}) & (Y_b + Y_c) \end{bmatrix} \begin{bmatrix} V_{an} \\ V_{ng} \end{bmatrix}.$$

Invert the admittance matrix and rewrite the equations as:

$$\begin{bmatrix} V_{an} \\ V_{ng} \end{bmatrix} = \begin{bmatrix} Y_{aa} & Y_a \\ (Y_{ba} + Y_{ca}) & (Y_b + Y_c) \end{bmatrix}^{-1} \begin{bmatrix} IP_a \\ IP_b + IP_c \end{bmatrix}.$$

Then, the four necessary unknowns are provided for solving a b-c line-to-line fault.

For the line-to-line to ground fault (assume b-c-g), the characteristic under this condition is:

$$V_{bn} = V_{cn} = V_{ng} = 0$$

$$I_a = 0$$

and we get:

$$I_a = IP_a - Y_{aa}V_{an} = 0.$$

Therefore,

$$V_{an} = \frac{IP_a}{Y_{aa}}.$$

5.1.2.3 Single-Phase Fault

The characteristic for the single-phase line-to-ground fault (assume a-g) is:

$$V_{an} = V_{ng} = 0$$

and

$$I_b = I_c = 0.$$

Substitute this into the currents equations to yield:

$$I_b = IP_b - Y_{bb}V_{bn} - Y_{bc}V_{cn}$$

$$I_c = IP_c - Y_{cb}V_{bn} - Y_{cc}V_{cn}.$$

Solving these two equations for V_{bn} and V_{cn} gives:

$$\begin{bmatrix} V_{bn} \\ V_{cn} \end{bmatrix} = \begin{bmatrix} Y_{bb} & Y_{bc} \\ Y_{cb} & Y_{cc} \end{bmatrix}^{-1} \begin{bmatrix} IP_b \\ IP_c \end{bmatrix}.$$

5.1.2.4 Example

Make the topology in Fig. 5.1 as an example to do the short circuit analysis. First, convert the graph into an adjacency matrix as:

$$M_a = \begin{array}{c} \\ v_1 \\ v_2 \\ v_3 \\ v_4 \\ v_5 \\ v_6 \end{array} \begin{pmatrix} \begin{array}{cccccc} v_1 & v_2 & v_3 & v_4 & v_5 & v_6 \\ 0 & 1 & 0 & 0 & 0 & 0 \\ 1 & 0 & 1 & 1 & 1 & 0 \\ 0 & 1 & 0 & 0 & 0 & 0 \\ 0 & 1 & 0 & 0 & 0 & 1 \\ 0 & 1 & 0 & 0 & 0 & 0 \\ 0 & 0 & 0 & 1 & 0 & 0 \end{array} \end{pmatrix}.$$

Then, extract relevant data from the preset data structure in Section 4.5, Chapter 4. Calculate the branch impedance, construct the A, B, c, and d matrix for the transformer, generate the capacitor unit, and calculate all load currents. Assume a grounded fault occurred on the segment between node 2 and node 3 as shown in Fig. 5.3; the new topology with the corresponding adjacency matrix is:

$$M_a = \begin{array}{c} \\ v_1 \\ v_2 \\ v_3 \\ v_4 \\ v_5 \\ v_6 \\ v_7 \end{array} \begin{array}{c} \begin{array}{ccccccc} v_1 & v_2 & v_3 & v_4 & v_5 & v_6 & v_7 \end{array} \\ \left(\begin{array}{ccccccc} 0 & 1 & 0 & 0 & 0 & 0 & 0 \\ 1 & 0 & 0 & 1 & 1 & 0 & 1 \\ 0 & 0 & 0 & 0 & 0 & 0 & 1 \\ 0 & 1 & 0 & 0 & 0 & 1 & 0 \\ 0 & 1 & 0 & 0 & 0 & 0 & 0 \\ 0 & 0 & 0 & 1 & 0 & 0 & 0 \\ 0 & 1 & 1 & 0 & 0 & 0 & 0 \end{array} \right) \end{array}.$$

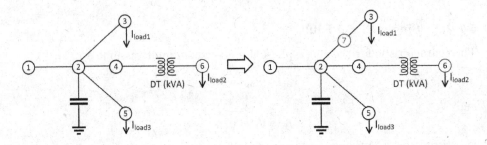

FIGURE 5.3
Short circuit analysis example topology of a radial system.

The line distance of each line segment needs to be set, and in this case they are set as 1500 ft, 2000 ft, 2500 ft, 1800 ft, and 500 ft, separately. Therefore, the following script can be used to calculate the short circuit currents and voltages under different phase(s) fault. In this situation, the fault node number is 7.

Listing 5.1
Short Circuit Analysis

```
1   % Presume the neutral is grounded.
2
3   e = input('the fault node number is: ');
4   n = parents(AdjM,e);
5   schild = children(AdjM,e);
6   f = [40 0 0
7        0 40 0
8        0 0 40];
9   Yf = (B{e}+f)^(-1);
10
11  ind = input('the number of fault phase is: ');
```

```
12
13   if ind == 3
14       V{e} = zeros(3,1);
15       I{e} = Yf * V{e};
16
17       for i =1:length(schild)
18       I{schild(i)} = Yf * V{e};
19       V{schild(i)} = zeros(3,1);
20       end
21
22   elseif ind == 2
23
24       temp = Yf * V{n};
25       I{e}(2) = temp(2) - V{n}(2)*((B{e}(1,2))^(-1));
26       I{e}(3) = temp(3) - V{n}(3)*((B{e}(1,3))^(-1));
27       I{e}(1) = 0;
28       V{e}(2) = 0; V{e}(3) = 0; V{e}(1) = temp(1)./B{e}(1);
29
30       temp1 = Yf * V{e};
31       for i =1:length(schild)
32       I{schild(i)}(2) = temp1(2) - ...
             V{e}(2)*((B{schild(i)}(1,2))^(-1));
33       I{schild(i)}(3) = temp1(3) - ...
             V{e}(3)*((B{schild(i)}(1,3))^(-1));
34       I{schild(i)}(1) = 0;
35       V{schild(i)}(2) = 0; V{schild(i)}(3) = 0;
36       V{schild(i)}(1) = temp1(1)./B{schild(i)}(1);
37       end
38
39   else
40       temp = Yf * V{n};
41       I{e}(1) = temp(1);
42       I{e}(2) = 0;
43       I{e}(3) = 0;
44       V{e}(1) = 0; V{e}(2) = temp(2)./B{e}(2); V{e}(3) = ...
             temp(3)./B{e}(3);
45
46       temp1 = Yf * V{e};
47       for i =1:length(schild)
48       I{schild(i)}(1) = temp1(1);
49       I{schild(i)}(2) = 0;
50       I{schild(i)}(3) = 0;
51       V{schild(i)}(1) = 0;
52       V{schild(i)}(2) = temp1(2)./B{schild(i)}(2);
53       V{schild(i)}(3) = temp1(3)./B{schild(i)}(3);
54       end
55   end
```

5.2 Availability of Fault Indicators on Pole-Mounted Switches

Fault indicators (FIs) are utilized to assist utility management or operation to localize the faulted components on a distribution system. The search time needed to locate a fault can be significantly reduced with the application of FIs. The typical fault localization technique involves the operation of fuses, terminators, and/or multiple types of switches to detect if a particular section of the feeder or parts of components has been isolated. A progressive deterioration of mechanical components and insulation parts will be caused by each fuse operation or fault closure. Therefore, fault indicators can also reduce the number of fuses or switch operations, to extend the life of the whole system and guarantee high searching efficiency [136–139].

In power supply service, customers can usually accept and understand a temporary power outage if service can be restored promptly. If the outage duration becomes excessive, a lot of complaints will be reported to the customer service center. In most power outage cases, once the fault is located, service of the "healthy" section in the feeder can be restored rapidly by closing normally open (NO) connections or switching to alternate feeders.

As the search for the exact fault location can be time-consuming, the effective use of time requires coordination of crew team with the control center where the dispatchers might have the most up-to-date information from the SCADA and trouble call tickets from customer billing centers. Having well-established coordination between these entities will reduce the time needed to locate the fault, sectionalize faults to the minimum, repair the segment, and restore power back to normal. The total duration adds up if the uncertainty of location remains. The automation of fault search by FIs is connected to the application with a topological update from the SCADA system conditioned such that the topology is updated regularly and the dispatchers are able to harness all telemetered measurements, both analog and binary. The assessment of potential outage areas is based on all these inputs and coordination with crew members at sites [136]:

- on the system protected by current limiting fuses, or

- on the system with cables in close proximity to each other (as could occur in transformers or at some junction nodes on a feeder).

If the indicator fails to reset after the system is repaired, they are of no use to the system operators. The indicators available can be broken down into four broad categories, based on the mechanism used to reset the indicator. These resetting indicators can be in the forms of [136]:

- manual reset,

- at medium-voltage,

- at low-voltage,

- reset by current measurement.

These categories of fault indicators can be coordinated within a feeder by various items, such as the location and its sensitivity.

Manually resetting fault indicators is a low-cost device that is often used on shielded or unshielded conductors or on the underground cable beneath the concentric neutral. The fault current that exceeds its trip rating generates adequate magnetic flux to rotate the permanent magnet while it cannot be tripped manually. To reset the indicators, a battery-operated tool or built-in button is set up with a button for that purpose.

Resetting indicators for a medium-voltage network are reset by the voltage of the cable they are mounted to. It is often deployed on overhead lines where it is utilized for that purpose as a potential fault indication. When a fault is cleared, the counter is re-initiated and is prepared for the next event. The pole-mounted FRTU has this feature and it is conveniently reachable for the crew team for maintenance. The counter is ideally suited to lock out a recloser, as well as current-limiting and expulsion-fuse-protected systems. This device provides a means of effective fault identification, but, it cannot be tripped itself because relay needs to be equipped with this device in order to trip.

Resetting indicators for a low voltage network are designed for underground systems where test points are not used. The device uses the same mechanism but it is placed under the concentric neutral directly under each terminator. It has electromagnetically tripped features where the resetting is accomplished by the secondary voltage of a distribution transformer and cannot be tripped mechanically.

The current resetting indicators are one of the largest family of fault indicators. There is large range of fault indicator options, e.g., single-phase trip with reset, three-phase trip on a single-phase fault with either single-phase or three-phase reset. The status checker is available locally and remotely. These devices are tripped electromechanically and cannot be tripped mechanically.

5.2.1 Placement of Fault Indicators (FIs)

The FI informs the repairing or switching crew while a fault occurs in its downstream network. As defining optimal FI placement is imperative for distribution utilities, FI placement is formulated as a power system optimization problem. The aims of prevalent FIs placement problems are to find the optimal number and locations of FIs, while the technical constraints are respected with minimum cost of deployment in the long term.

Since the number of affected customers and the outage duration are the main factors that need to be considered in evaluating the service quality and reliability for a distribution system, the data logging of these two factors is generated for each fault event. In the statistical analysis of service outage, the customer interruption cost (CIC) C_{INT} of a distribution system can be expressed as follows [140]:

$$C_{\text{INT}} = \sum_{i=1}^{n_l} CY_{\text{INT},i} = \sum_{i=1}^{n_l} r_{o,i} \cdot l_i \left(\sum_{j=1}^{n_l} C_{\text{INT},ij} \cdot L_j \right)$$

where

n_l is the total number of line segments;

$CY_{\text{INT},i}$ is the interruption cost per year due to outages in line segment i;

$r_{o,i}$ is the outage rate (failure per year/mile) of line segment i;

l_i is the length of line segment i;

$C_{\text{INT},ij}$ is the interruption cost of load at segment j due to an outage at segment i;

L_j is the total load of line segment j.

The C_{ij} represents the integrated interruption costs generated from different types of customers (residential, commercial, and/or industrial). Therefore, the service reliability can be improved by placing FIs in the proper positions to reduce the outage losses. To achieve this purpose, each FI has to be installed at a zonal detectable location and/or lateral levels within the feeder. Then, the total cost of reliability (TCR) to be minimized is defined as [140]:

$$\text{Min TCR} = \text{CIC} + \text{INVC}$$

where INVC is the investment cost of FIs.

Similarly, the total customer interruption durations (CID) D_{INT} can be represented as:

$$D_{\text{INT}} = \sum_{k=1}^{n_c} DY_{\text{INT},k} = \sum_{k=1}^{P} r_{o,k} \cdot n_{c,k} \left(\sum_{g=1}^{P} D_{k,g} \cdot n_{c,g} \right)$$

where

n_c is the total number of customers;

$DY_{\text{INT},k}$ is the interruption duration of customer k per year due to outages;

P is the total number of load points;

$r_{o,k}$ is the outage rate of customer k;

$D_{k,g}$ is the interruption duration of load point g due to an outage at load point k;

$n_{c,g}$ is the total number of customers of load point g.

The total interruption duration (TID) to be minimized is defined as:

$$\text{Min TID} = \text{CID} + \text{REST}$$

where REST is the customer restoration time. To reduce the search space and provide the optimal installation plan, the following FI installation rules are proposed [140–142]:

1. The feeders or laterals serving high energy-consumption loads with relatively high power-outage frequency, such as high technology plants, industrial areas, and business metropolises, have higher priority to install FIs.

2. The FI installation environment is divided into two types: 1) three-phase, four-way sectionalizing cabinets, on non-automatic switches, or 2) three-phase, feed-through, pad-mounted transformers.

3. It is not necessary to install the FIs on the normally open (NO) tie switches.

5.2.2 Fault Segment Localization

The objectives of fault segment localization are to minimize the interruption of power delivery and to facilitate power restoration in a timely manner, which will improve overall system reliability. The fault indicators (FI) in the computerized automation for an energy SCADA network provide a clue where the faulted location may fall in an area. This will help a crew who navigates the area determine the exact location, and electrically connect the unaffected customers by temporarily connecting the NO switches from another feeder. This effort of localization is often centralized in the DMS system where topology information and all states of binary control and analog variables are updated by SCADA systems. There can be either momentary or permanent faults that can be harnessed by the DMS system in order to provide the best recommendation to dispatchers and the crew who is actively searching for an exact fault location. The initial investigation of a fault can be subsequent to sectionalizing the non-remote-controlled switches that minimize the outage area if a permanent fault is detected. The generation of switching steps will be discussed in a later chapter. Momentary faults will be recorded with probable function by tracing the alarms and tripped sequences sent to the SCADA system.

Under common circumstances, the following operations can help crews to determine the probable fault location(s) in the system [143]:

1. A temporary fault has occurred and no fault location information can be inferred from this operation sequence if one recloser trips (open) followed by a successful reclose in 1 to 3 seconds (i.e., no fault current on the first reclose).

2. A fault most likely caused a fuse burn if one recloser closes and measures fault current but does not trip the recloser on the time delay curve. The fault current remains on the first reclose but there are no operations except the first trip.

3. A fault was most likely beyond a line sectionalizer, and the sectionalizer is open if three times of the recloser trip, but there is no fault current on the third (last) recloser close operation.

4. The most probable location of a fault is on the main feeder line or unprotected laterals if four times of the recloser trip (lock out).

Prediction of fault distance from the feeder head is based on the magnitude of fault current at the feeder head where measurements can be archived and utilized for analysis. Sensitive analysis with varying fault impedance values can determine the range of distance using established circuit methods. One of the key requirements to get the benefits of automating search of fault in SCADA is the FRTU control functionality. Most units today are equipped with fault status in which all FRTUs are an integral part of SCADA. The search time would depend on the number of deployed FRTUs, which is an upgrade from the existing sectionalizers. A feeder may consist of many of them and typically the upgrade may be a small fraction for each feeder. However, the ideal case is that all sectionalizers for a feeder may eventually be upgraded with FRTUs that enable many possibilities of reconfiguration. These remote-controlled switches (RCSs) may have fault indicators, but they do not play the role of breakers or reclosers. The binary information of the FI sends the latest status to the dispatching control center (DCC). Although these FIs can provide a decent hint for the search of the exact location, it is often tricky to be conclusive as the timeliness and latency of the SCADA network may cause delays or losses of incoming messages from the remote sites as well as wait time for the fault localization module to make an inference. Typically, the FI can provide a digital binary measurement (Yes or No). However, when the communication infrastructure of the radial network is weak, the management panel may not receive the corresponding data in a timely manner. Some disturbances may affect the communication, which may result in "unknown" status for the FI.

In the fault localization analysis, the FI only considered "Yes" or "No." A feeder with its visual representation and the equivalent graph representation is illustrated in Fig. 5.4. The empty box here indicates the tripped breaker and two remote-controlled switches with "Yes" and "No" FIs. Each area in this case actually consists of multiple nodes and edges, and the switch is the two-terminal unit shown as an edge in the graph representation. Since the "Yes" indicates the following feeder has a fault and "No" represents normal, the fault exists in area 2 in this example.

The remote-controlled switch does not have the relay such that when the switch is tripped, it is actually tripped over the high-end but most of the switches remain closed. Differing from the previous example, in the more com-

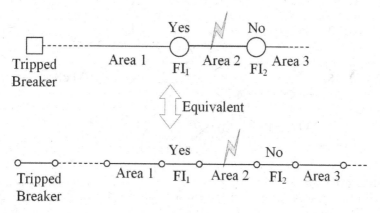

FIGURE 5.4
Feeder visual representation and the equivalent graph representation with fault indicators.

plicated topology shown in Fig. 5.5, the switches are represented by vertices. The possible fault area is between the "Yes" and "No" FIs and indicated by the red zone (see color e-book). An extended analysis based on the same topology to discuss the location of the fault segment with different indications on the "question mark" FI is demonstrated in Fig. 5.6. The following are the possibilities:

- If "?" is "Yes": the fault segment exists in Area 2;

- If "?" is "No": the fault segment exists in Area 1;

- If "?" is "Unknown": the crew needs to search the union of both areas, 1 and 2.

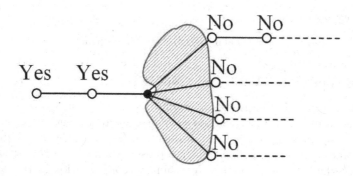

FIGURE 5.5
Fault area localization with fault indicators.

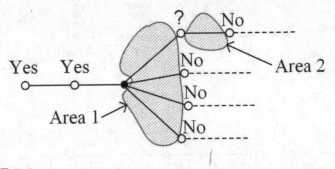

FIGURE 5.6
Fault area localization with fault indicators.

5.3 Ideal Scenario — All Switches Are Remote-Controlled

The ideal scenarios demonstrated here are not even close to real-world topology, but they are close enough to realize the graph-based fault localization concept.

The assumptions on graph-based fault localization analysis here are:

- All nodes and edges are optimally re-ordered;

- Lines, loads, and switches, etc., are topologically connected;

- The subsystem is radially set up under normal conditions (spanning tree);

- A breaker is equipped with relay;

- Switches (disconnectors) are remote-controlled but without relay;

- Switches have FIs;

- No fuses are modeled for each distribution transformer.

5.3.1 Scenario 1: "Yes" and "No" FIs

As shown in Fig. 5.7, the graph representation consists of 8 vertices and 7 edges. Line segments 23, 46, 78, and 45 represent load areas. When a fault occurred in segment 46 (Area 2) or segment 45 (Area 4), the FI_1 indicated as "Yes" and the corresponding binary representation is 1 in this situation. Conversely, the FI_2 indicated as "No" (binary representation is 0) since the following segments in area 3 work normally. As a result, the entire feeder is de-energized, since the circuit breaker is tripped.

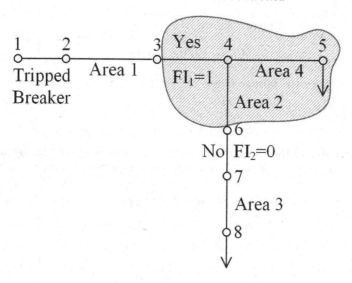

FIGURE 5.7
A scenario of the fault area localization.

Initialization

The initialization of the graph-based analysis applied to FIs needs two row vectors and the topology matrix. In this example, the incidence matrix is:

$$
M_i = \begin{array}{c} \\ v_1 \\ v_2 \\ v_3 \\ v_4 \\ v_5 \\ v_6 \\ v_7 \\ v_8 \end{array}
\begin{array}{cccccccc}
e_1 & e_2 & e_3 & e_4 & e_5 & e_6 & e_7 \\
\left(\begin{array}{ccccccc}
1 & 0 & 0 & 0 & 0 & 0 & 0 \\
1 & 1 & 0 & 0 & 0 & 0 & 0 \\
0 & 1 & 1 & 0 & 0 & 0 & 0 \\
0 & 0 & 1 & 1 & 1 & 0 & 0 \\
0 & 0 & 0 & 1 & 0 & 0 & 0 \\
0 & 0 & 1 & 0 & 1 & 1 & 0 \\
0 & 0 & 0 & 0 & 0 & 1 & 1 \\
0 & 0 & 0 & 0 & 0 & 0 & 1
\end{array}\right)
\end{array}.
$$

The FI_1 is e_3 and the FI_2 is e_6 in this example. The initial edge connection vector is:

$$E = [1\ 1\ 1\ 1\ 1\ 1\ 1].$$

However, $FI_1 = 1$ and $FI_2 = 0$, as mentioned, reverse the connection state so that the edge vector with FI information displayed in the corresponding position is:

$$E_{FI} = [1\ 1\ 0\ 1\ 1\ 1\ 1].$$

Furthermore, since the topology is de-energized, the initial vertex indication vector for the affected area is shown as:

$$V_D = [1\ 0\ 0\ 0\ 0\ 0\ 0\ 0]$$

where 1 indicates the power source.

Step 1

The first step of the analysis is to convert the incidence matrix to the adjacency matrix, and include the latest FIs' information to update the topology status:

$$\widehat{M_i} = M_i \times \text{diag}(E_{\overline{FI}}).$$

$$\widehat{M_a} = \widehat{M_i} \times \widehat{M_i}^\top.$$

$$M_a = \left| \widehat{M_a} - \text{diag}(\text{diag}(\widehat{M_a})) \right|.$$

The generated adjacency matrix is:

$$M_a = \begin{array}{c} \\ v_1 \\ v_2 \\ v_3 \\ v_4 \\ v_5 \\ v_6 \\ v_7 \\ v_8 \end{array} \begin{array}{c} \begin{array}{cccccccc} v_1 & v_2 & v_3 & v_4 & v_5 & v_6 & v_7 & v_8 \end{array} \\ \left(\begin{array}{cccccccc} 0 & 1 & 0 & 0 & 0 & 0 & 0 & 0 \\ 1 & 0 & 1 & 0 & 0 & 0 & 0 & 0 \\ 0 & 1 & 0 & 0 & 0 & 0 & 0 & 0 \\ 0 & 0 & 0 & 0 & 1 & 1 & 0 & 0 \\ 0 & 0 & 0 & 1 & 0 & 0 & 0 & 0 \\ 0 & 0 & 0 & 1 & 0 & 0 & 1 & 0 \\ 0 & 0 & 0 & 0 & 0 & 1 & 0 & 1 \\ 0 & 0 & 0 & 0 & 0 & 0 & 1 & 0 \end{array} \right) \end{array}.$$

Step 2

The second step is an iterative step, which is to find the non-affected nodes connected with the previous node.

$$V_D' = V_D \times \widehat{M_a} + V_D = [1\ 1\ 0\ 0\ 0\ 0\ 0\ 0].$$

$$V_D'' = V_D' \times \widehat{M_a} + V_D' = [2\ 2\ 1\ 0\ 0\ 0\ 0\ 0].$$

Convert all non-zero elements in this vector to 1 so that:

$$V_D'' = [1\ 1\ 1\ 0\ 0\ 0\ 0\ 0].$$

Keep the iteration until the generated vector is the same as in the previous step, then

$$V_D^{\text{new}} = [1\ 1\ 1\ 0\ 0\ 0\ 0\ 0].$$

In this example, the nodes 4, 5, 6, 7, 8 are affected, since the corresponding elements in this vector are zeros (de-energized).

Step 3

The third step is to search line segments based on the incidence matrix, to localize the possible fault area. In this case, because E_6 is an FI and it shows normally, the search will start from the first edge connected to affected nodes untill E_6. It should be noted that all vertices and edges have been topologically ordered. As the search result shown on the incidence matrix, the possible fault area, in this case, includes segments 3, 4, and 5, which indicate FI_1, area 2, and area 4 in Fig. 5.7.

$$
M_i =
\begin{array}{c}
\\
v_1 \\
v_2 \\
v_3 \\
v_4 \\
v_5 \\
v_6 \\
v_7 \\
v_8
\end{array}
\begin{array}{ccccccc}
e_1 & e_2 & e_3 & e_4 & e_5 & e_6 & e_7 \\
1 & 0 & 0 & 0 & 0 & 0 & 0 \\
1 & 1 & 0 & 0 & 0 & 0 & 0 \\
0 & 1 & 1 & 0 & 0 & 0 & 0 \\
0 & 0 & 1 & 1 & 1 & 0 & 0 \\
0 & 0 & 0 & 1 & 0 & 0 & 0 \\
0 & 0 & 1 & 0 & 1 & 1 & 0 \\
0 & 0 & 0 & 0 & 0 & 1 & 1 \\
0 & 0 & 0 & 0 & 0 & 0 & 1
\end{array}
.
$$

5.3.2   Scenario 2: "Yes" and "Yes" FIs

As shown in Fig. 5.8, the graph representation applies the same topology as scenario 1, but both the FI_1 and FI_2 are equal to 1, and are indicated as "Yes" in this case. Also, the entire feeder is de-energized, since the circuit breaker is tripped.

Initialization

In this example, the initial incidence matrix is the same as in the previous case, while the edge vector with FIs information is:

$$E_{\overline{FI}} = [1\ 1\ 0\ 1\ 1\ 0\ 1],$$

since FI_1 and FI_2 are represented as edges 3 and 6, separately. The initial vertex indication vector is shown as:

$$V_D = [1\ 0\ 0\ 0\ 0\ 0\ 0\ 0].$$

Step 1

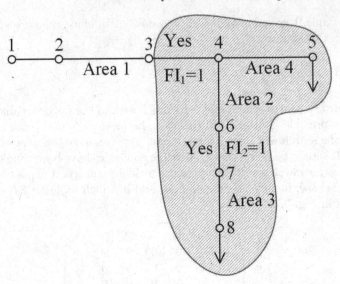

FIGURE 5.8
A scenario of the fault area localization.

The converted adjacency matrix with the updated topology information
is:

$$
M_a = \begin{array}{c} \\ v_1 \\ v_2 \\ v_3 \\ v_4 \\ v_5 \\ v_6 \\ v_7 \\ v_8 \end{array}
\begin{array}{c} \begin{array}{cccccccc} v_1 & v_2 & v_3 & v_4 & v_5 & v_6 & v_7 & v_8 \end{array} \\
\left(\begin{array}{cccccccc}
0 & 1 & 0 & 0 & 0 & 0 & 0 & 0 \\
1 & 0 & 1 & 0 & 0 & 0 & 0 & 0 \\
0 & 1 & 0 & 0 & 0 & 0 & 0 & 0 \\
0 & 0 & 0 & 0 & 1 & 1 & 0 & 0 \\
0 & 0 & 0 & 1 & 0 & 0 & 0 & 0 \\
0 & 0 & 0 & 1 & 0 & 0 & 0 & 0 \\
0 & 0 & 0 & 0 & 0 & 0 & 0 & 1 \\
0 & 0 & 0 & 0 & 0 & 0 & 1 & 0
\end{array} \right) \end{array}.
$$

Step 2

The generated nodes vector is:

$$V_D^{\text{new}} = [1\,1\,1\,0\,0\,0\,0\,0]$$

which indicates the nodes 4, 5, 6, 7, 8 are affected (de-energized).

Step 3

The search result in this step is shown on the incidence matrix as:

$$M_i = \begin{array}{c} \\ v_1 \\ v_2 \\ v_3 \\ v_4 \\ v_5 \\ v_6 \\ v_7 \\ v_8 \end{array} \begin{array}{ccccccc} e_1 & e_2 & e_3 & e_4 & e_5 & e_6 & e_7 \\ \left(\begin{array}{ccccccc} 1 & 0 & 0 & 0 & 0 & 0 & 0 \\ 1 & 1 & 0 & 0 & 0 & 0 & 0 \\ 0 & 1 & 1 & 0 & 0 & 0 & 0 \\ 0 & 0 & 1 & 1 & 1 & 0 & 0 \\ 0 & 0 & 0 & 1 & 0 & 0 & 0 \\ 0 & 0 & 1 & 0 & 1 & 1 & 0 \\ 0 & 0 & 0 & 0 & 0 & 1 & 1 \\ 0 & 0 & 0 & 0 & 0 & 0 & 1 \end{array}\right) \end{array}$$

which indicates the possible fault area includes edges 3, 4, 5, 6 and 7 as shown in Fig. 5.8.

5.3.3 Scenario 3 (Rare Case): "No" and "Yes" FIs

This scenario is a rare case only if the feeder head FI supervises a limited segment or the FI is not sensitive. As shown in Fig. 5.9, the graph representation also applies the same topology as in scenario 1, but the $FI_1 = 0$, which indicated as "No," and the $FI_2 = 1$ to represent "Yes" in this case. Furthermore, the entire feeder is de-energized, since the circuit breaker is tripped.

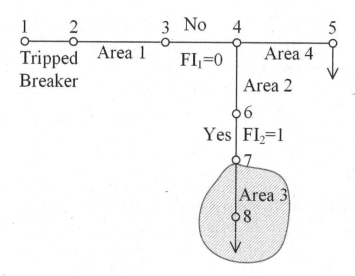

FIGURE 5.9
A scenario of the fault area localization.

Initialization

In this example, the initial incidence matrix is the same as in scenario 1, while the edge vector with FIs information is:

$$E_{\overline{FI}} = [1\ 1\ 1\ 1\ 1\ 0\ 1]$$

since $FI_2 = 1$ (edge 6) this time. The initial vertex indication vector is shown as:

$$V_D = [1\ 0\ 0\ 0\ 0\ 0\ 0\ 0].$$

Step 1

The converted adjacency matrix with the updated topology information is:

$$M_a = \begin{array}{c} \\ v_1 \\ v_2 \\ v_3 \\ v_4 \\ v_5 \\ v_6 \\ v_7 \\ v_8 \end{array}\begin{pmatrix} \begin{array}{cccccccc} v_1 & v_2 & v_3 & v_4 & v_5 & v_6 & v_7 & v_8 \\ 0 & 1 & 0 & 0 & 0 & 0 & 0 & 0 \\ 1 & 0 & 1 & 0 & 0 & 0 & 0 & 0 \\ 0 & 1 & 0 & 1 & 0 & 0 & 0 & 0 \\ 0 & 0 & 1 & 0 & 1 & 1 & 0 & 0 \\ 0 & 0 & 0 & 1 & 0 & 0 & 0 & 0 \\ 0 & 0 & 0 & 1 & 0 & 0 & 0 & 0 \\ 0 & 0 & 0 & 0 & 0 & 0 & 0 & 1 \\ 0 & 0 & 0 & 0 & 0 & 0 & 1 & 0 \end{array} \end{pmatrix}.$$

Step 2

The generated nodes vector is:

$$V_D^{new} = [1\ 1\ 1\ 1\ 1\ 1\ 0\ 0]$$

which indicate nodes 7 and 8 are affected (de-energized) with the suggested fault area.

Step 3

The search result in this step is shown on the incidence matrix as:

$$M_i = \begin{array}{c} \\ v_1 \\ v_2 \\ v_3 \\ v_4 \\ v_5 \\ v_6 \\ v_7 \\ v_8 \end{array}\begin{pmatrix} \begin{array}{ccccccc} e_1 & e_2 & e_3 & e_4 & e_5 & e_6 & e_7 \\ 1 & 0 & 0 & 0 & 0 & 0 & 0 \\ 1 & 1 & 0 & 0 & 0 & 0 & 0 \\ 0 & 1 & 1 & 0 & 0 & 0 & 0 \\ 0 & 0 & 1 & 1 & 1 & 0 & 0 \\ 0 & 0 & 0 & 1 & 0 & 0 & 0 \\ 0 & 0 & 1 & 0 & 1 & 1 & 0 \\ 0 & 0 & 0 & 0 & 0 & 1 & 1 \\ 0 & 0 & 0 & 0 & 0 & 0 & 1 \end{array} \end{pmatrix}$$

which indicates the possible fault area includes edges 6 and 7. Then the fault exists in FI_2 or area 3 in Fig. 5.9. Listings 5.2 and 5.3 are the MATLAB codes for these cases. Listing 5.3 stores all these cases.

Listing 5.2
Fault Localization Analysis

```
1   function edge_state = floc(incMat, EdgeVec, NodeVec)
2
3   % Convert incidence matrix to adjacency matrix.
4   YesNoMat = diag(EdgeVec);
5   incMatr = incMat * YesNoMat;
6   adjMat = incMatr*incMatr';
7   adjMat = abs(adjMat - diag(diag(adjMat)));
8
9   % Initialize values.
10  affectnode_vec = NodeVec;
11  affectnode_old = affectnode_vec*0;
12
13  % Criteria that if the result is equal to the previous ...
        iteration result.
14  while (length(find(affectnode_old == 0)) - length(find ...
        (affectnode_vec == 0))) ~= 0
15      affectnode_old = affectnode_vec;
16      % Multiplication based on the topology (Adjacency Matrix).
17      affectnode_vec = affectnode_old * adjMat + affectnode_old;
18      idx = find(affectnode_vec > 0);
19      affectnode_vec(idx) = 1; % Replace all non-zero elements ...
            to 1s.
20  end
21
22  % Find affected connections based on affected nodes from ...
        incidence matrix.
23  line_state = ones(1,length(incMat(1,:)));
24  node_ind = find(affectnode_vec == 0);
25  for i = 1:length(node_ind)
26      j = node_ind(i);
27      line_ind = find(incMat(j,:) == 1);
28      line_state(line_ind) = 0;
29  end
30
31  % Consider 'NO' FIs in sub-feeder.
32  edge_state = line_state;
33  FIn = input('input the index of no FI: ');
34  for i = 1:length(FIn)
35      j = FIn(i);
36      node_subFIn = find(incMat(:,j)==1,1,'last');
37      edge_subFIn = find(incMat(node_subFIn,:)==1);
38
39      for x = 1:length(edge_subFIn)
40          y = edge_subFIn(x);
41          edge_state(y) = 1;
42      end
43  end
```

Listing 5.3
Fault Localization Case Study

```
1    casenumb = input('Input the case number (1-6): ');
2
3    % Case 1: Tree Topology: FI1-Yes, FI2-5-No
4    if casenumb == 1
5       IncM = [1 0 0 0 0 0 0 0 0 0 0 0
6              1 1 0 0 0 0 0 0 0 0 0 0
7              0 1 1 0 0 0 0 0 0 0 0 0
8              0 0 1 1 1 1 0 0 0 0 0 0
9              0 0 0 1 0 0 1 0 0 0 0 0
10             0 0 0 0 0 0 1 0 0 1 0 0
11             0 0 0 0 0 0 0 0 0 1 1
12             0 0 0 0 0 0 0 0 0 0 1
13             0 0 0 0 1 0 0 1 0 0 0
14             0 0 0 0 0 0 0 1 0 0 0
15             0 0 0 0 0 1 0 0 1 0 0
16             0 0 0 0 0 0 0 0 1 0 0];
17
18      E_D = [1 1 0 1 1 1 1 1 1 1 1 1];
19      V_D = [1 0 0 0 0 0 0 0 0 0 0 0];
20
21   % Case 2: Tree Topology: FI1-Yes, FI2-Yes, FI3-5-No
22   elseif casenumb == 2
23      IncM = [1 0 0 0 0 0 0 0 0 0 0 0
24             1 1 0 0 0 0 0 0 0 0 0 0
25             0 1 1 0 0 0 0 0 0 0 0 0
26             0 0 1 1 1 1 0 0 0 0 0 0
27             0 0 0 1 0 0 1 0 0 0 0 0
28             0 0 0 0 0 0 1 0 0 1 0
29             0 0 0 0 0 0 0 0 1 1
30             0 0 0 0 0 0 0 0 0 1
31             0 0 0 0 1 0 0 1 0 0 0
32             0 0 0 0 0 0 0 1 0 0 0
33             0 0 0 0 0 1 0 0 1 0 0
34             0 0 0 0 0 0 0 0 1 0 0];
35
36      E_D = [1 1 1 1 1 1 0 1 1 1 1];
37      V_D = [1 0 0 0 0 0 0 0 0 0 0 0];
38
39   % Case 3: Scenario 1: 6-node, FI1-Yes, FI2-No
40   elseif casenumb == 3
41      IncM = [1 0 0 0 0 0 0
42             1 1 0 0 0 0 0
43             0 1 1 0 0 0 0
44             0 0 1 1 1 0 0
45             0 0 0 1 0 0 0
46             0 0 0 0 1 1 0
47             0 0 0 0 0 1 1
48             0 0 0 0 0 0 1];
49
50      E_D = [1 1 0 1 1 1 1];
51      V_D = [1 0 0 0 0 0 0];
52
53   % Case 4: Scenario 2: 6-node, FI1-No, FI2-Yes
54   elseif casenumb == 4
```

```
55      IncM = [1 0 0 0 0 0 0
56             1 1 0 0 0 0 0
57             0 1 1 0 0 0 0
58             0 0 1 1 1 0 0
59             0 0 0 1 0 0 0
60             0 0 0 0 1 1 0
61             0 0 0 0 0 1 1
62             0 0 0 0 0 0 1];
63
64      E_D = [1 1 1 1 1 0 1];
65      V_D = [1 0 0 0 0 0 0 0];
66
67  % Case 5: Scenario 3:6-node, FI1-Yes, FI2-Yes
68  elseif casenumb == 5
69      IncM = [1 0 0 0 0 0 0
70             1 1 0 0 0 0 0
71             0 1 1 0 0 0 0
72             0 0 1 1 1 0 0
73             0 0 0 1 0 0 0
74             0 0 0 0 1 1 0
75             0 0 0 0 0 1 1
76             0 0 0 0 0 0 1];
77
78      E_D = [1 1 1 1 1 0 1];
79      V_D = [1 0 0 0 0 0 0 0];
80
81  % Case 6: Scenario 4: extension of scenario 1, 10-node, ...
        FI1-Yes, FI2-No
82  elseif casenumb == 6
83      IncM = [1 0 0 0 0 0 0 0 0
84             1 1 0 0 0 0 0 0 0
85             0 1 1 0 0 0 0 0 0
86             0 0 1 1 1 0 0 0 0
87             0 0 0 1 0 0 0 1 0
88             0 0 0 0 1 1 0 0 0
89             0 0 0 0 0 1 1 0 0
90             0 0 0 0 0 0 1 0 0
91             0 0 0 0 0 0 0 1 1
92             0 0 0 0 0 0 0 0 1];
93
94      E_D = [1 1 0 1 1 1 1 1 1];
95      V_D = [1 0 0 0 0 0 0 0 0 0];
96  end
```

5.4 Conclusions

As a design in protective relaying, short circuit analysis is always presumed with a specific location of fault and fault type in order to determine the maximum fault current passing through a feeder. However, the reversal of that process is to search for a faulted segment, which can be challenging.

This can be inferred from the pole-mounted fault indicators. Although it may take time to locate an exact location, the protective relaying would quickly disconnect the faulted segment. This chapter explores a given graph based on the energization state from the protective tripping and latest statuses of fault indicators to pinpoint the faulted segment. This chapter will relate to the next chapter on restoration. It is essential that the smallest fault segment is identified before connecting to partial restoration, which will eventually lead to complete restoration once the fault segment is diagnosed. This chapter also addresses inconsistent fault indication, which requires a system to determine a larger search area by crew members.

Mini Project 4: Integration of Short Circuit and Fault Localization Modules

The previous chapters have provided MATLAB script to input datasets from GIS to graph (either incidence or adjacency matrices). Modify the given script in this chapter and integrate it with your script thus far. Upon completion of script integration, ensure that the script functions as it should behave. Your verification depends on how you establish the randomly generated fault on each segment of your line, defined as a "faulted" segment within a feeder that is reflected on the graph matrix. Note that the current version of the script will only handle a single event within a feeder. This is the perception that you must beware of when verifying the output.

6

Isolation of Faulted Segment and Partial Restoration

This chapter follows the previous description that continues the notion of automation. While the previous chapter indicates a large fault area has been concluded, the remainder of the healthy segment of the tripped feeder may remain de-energized. If the tripped feeder has connected with the neighboring remote-controlled normally open (NO) feeders, the isolation can be further improved by going through bisection search. The trial and error of opening the non-remote-controlled switches may trip and indicate if the fault is upstream or downstream. Several iterative attempts of such efforts can infer the whereabouts of the smallest segment that is felt within non-remote-controlled switches. Then, the power flow module will kick in to evaluate each scenario with results that recommend the optimal case with total power losses and voltage violation, if any. The remote-controlled NO switch will be opened by dispatchers from the control center. Such temporary reconfiguration is often very effective to reduce total outages and improve overall reliability [144,145].

To improve the reliability of the power supply and the service quality, the operators will first identify the faulted section and control based on the remote-controlled switches (RCSs) on the faulted feeder from the remote sites to isolate the faulted area. This can be a large de-energized area. The distribution system line models may consist of overhead and underground types. In the case of power outage due to a fault and/or a scheduled outage, tie switches transfer the fault-free existing outage area to other possible feeder(s). The reliability of the distribution system can be enhanced greatly by avoiding the need for crew patrols and manual switching operations [146].

Once a fault occurs in a feeder, the recloser will attempt to reclose up to three times. If it is a permanent fault, the protective device on the faulted segment will be opened in lockout status. Then, the DMS identifies the faulted area based on the fault-related information, which can be collected sequentially from RTUs/FRTUs and the fault indicators (FIs). After entirely isolating the faulted section, the protective device is closed back and the network reconfiguration strategy transfers the "healthy" outage zone to the alternative feeders. Due to multiple aspects of effects, such as the high impedance, inrush currents, and unbalanced loads, the fault is difficult to accurately determine. The fundamental rules are that if the phase current is greater than the defined overcurrent set as the relay pick-up value, then the FI will be set to on to raise the red flag that this fault might occur here. On the other hand, if the fault is beyond the faulted segment downstream of the feeder, then the FI will be set off and indicates in the SCADA system. This can occur when a high impedance fault occurs. Under this circumstance, the FI may not show up because the phase current is smaller than the pick-up value. Under an abnormal condition, the status of the FI may show when the load current exceeds the overcurrent relay pick-up value due to the inrush characteristics [147,148]. Upon a reclosing lockout, the load area under the protective device for the fault zone experiences outage and can only be tried again from the centralized SCADA using the remote control function recommended by the DMS applications.

6.1 Operating States

The operating state of a distribution feeder is bound under two sets of constraints, i.e., (1) load constraints and (2) operating constraints. In the SCADA network, the FRTU serves as the root nodes of all customers associated with a subsystem. The lumped load associated with the FRTUs (upstream or downstream, if any) shall operate within the acceptable limits, e.g., maximum primary feeder current and voltage regulation. Violations in normal-condition feeder current limits can be tolerated for some time (short-term limits), whereas emergency limits usually are avoided due to proper overcurrent protection schemes [144].

As shown in Fig. 6.1, the system states can be categorized into 3 major states, i.e., *normal, emergency,* and *restorative.* A system is in normal state when there is no violation of emergency limits while satisfying all loads in the system. An emergency state occurs when any of the feeders experiences a power outage due to a fault and some operating limtis are violated. Finally, a restorative state is when a remedial action is taken from the emergency state and a fault is successfully isolated. This effort is initiated by the operators in the control room and reconfiguration is temporarily arranged to energize power to the healthy part of the feeder segment from a neighborhood NO switch.

Fig. 6.1 also illustrates some of the possible transitions between different operating states. Some state transitions are involuntary, since they are not controlled by the management system. These operations include protective control actions, such as the response of faults and load variations. This section presents the strategies of voluntary state transitions in the distribution system by centralized control [144].

- **Normal to Normal**: Under normal state, all the operating limits and constraints are satisfied, and the main purpose of operations is to improve

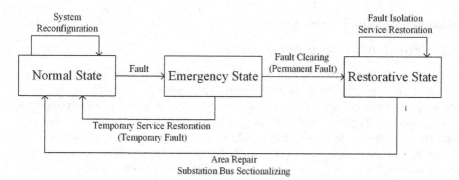

FIGURE 6.1

Operating states of a distribution system [144].

the service quality by optimizing the voltage profile, minimizing power losses, and/or balancing the feeder loads. Therefore, the objective, in this case, is to find a feasible operating point in the sense that normal operating limits are also satisfied.

- **Normal to Emergency**: The fault current with a high magnitude on a feeder caused by a fault or short circuit occurs on the primary network. The system state transits from normal to emergency. Operations under this situation should be quickly cleared to avoid equipment or component damage.

- **Emergency to Normal**: If the fault is temporary and there is no fault current on the first reclose of a recloser, the operator can reclose the tripped circuit breaker, which takes the system back to the normal state.

- **Emergency to Restorative**: The fault clearing and repairing actions will be performed if a permanent fault occurs on a feeder. This transition takes the system to the restorative state where all the loads on the faulted feeder have been disconnected (i.e., all customers in the faulted area are out of service).

- **Restorative to Restorative**: During the fault clearance, the fault localization and isolation operations will be performed to identify and isolate the faulted area, and then, temporary partial service restoration for the "healthy" load in the isolated zone may be carried out by breaker/recloser operations from the adjacent feeders or substations. The system remains in the restorative state, while the fault is not repaired.

- **Restorative to Normal**: After the fault clearance, the faulted area is re-energized. In addition, if the fault is located in the distribution substation, the substation bus sectionalizing is carried out. Without repairing the fault, the distribution system can also be taken back to the normal state.

6.2   Fault Management System

In terms of the interrelationship between each operation and process within the fault management system, the disturbance generates a fault in the line segment on a feeder so that the circuit breaker on the feeder head is tripped. Then the DMS performs the fault segment localization according to the information gathered from the fault indicators. After identifying the faulted segment, the automatic generation switching steps break the corresponding remote-controlled switches (sectionlizers) to isolate the fault. Usually, this operation will isolate a large area within a distribution system.

Once the faulted area is isolated, temporary service will be restored for the fault-free segments by reclosing the tripped breaker and closing the tie switches from other feeders. If there is at least one tie switch, i.e., NO switch, connected with an adjacent feeder, the restoration is possible. This operation is topologically feasible, but it is not a feeder reconfiguration since there are limited switching options and operations. With the switching operations, the faulted segment can be narrowed down until the smallest area is found. Then the distribution control center will send a field crew to find and repair the problem. After a few hours of fixing, the whole system will be restored back to normal and the topology will be back to the initial state. The framework of the fault management system is demonstrated in Fig. 6.2. The fault management system can be applied to both MV- and LV-faults. However, the process is somewhat different for MV- and LV-faults.

6.2.1   MV Faults (Faults Occur on a Primary Network)

The MV faults are referred to the electrical short circuit event that occurs within the primary feeder, and the breaker or recloser(s) may de-energize a large area of the feeder. Consequently, hundreds of customers associated with the feeder experience power outage. It is a major priority for utilities to pinpoint the exact fault location. The fault event is initiated by a protective relay trips a circuit breaker with typically three attempts before locking out. Operators are informed with alarms on the SCADA system, including showing fault indicators on those RCSs. However, high-impedance earth faults may not be effectively detected by the relay protection. If the incoming FI binary statuses are reported to the SCADA system within the observed timeframe, then the one-line diagram showing on only switches in a feeder will highlight potential faulted segments. A well-trained operator will be able to begin the best guess to narrow down the search segment between non-remote-controlled switches after several trial-and-errors on opening other RCS(s) recommended by the DMS fault inference application.

The isolation of a fault can be a straightforward process if all events arrive in the SCADA system where topological statuses are updated from the

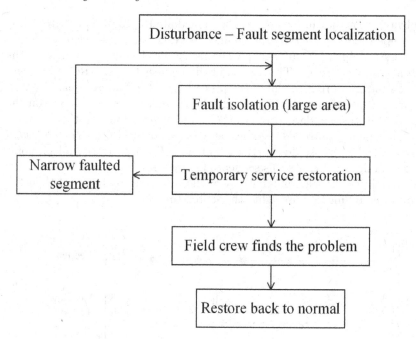

FIGURE 6.2
The framework of the fault management system.

FRTUs/RTUs. Once the fault is isolated (larger area), the crew members are dispatched to the site to trace along the "problematic" feeder segment. Visually, it can be a non-trivial task to find an exact location, and so the crew might coordinate with the operators in the control center to coordinate opening on non-remote-controlled switches, i.e., manually open any disconnectors iteratively until the smallest segment is isolated. Experimental switching upstream or downstream is performed with careful switching and coordination between the two parties to locate the faulted zone by opening disconnectors along the feeder. Each try will involve closing the circuit breaker to observe if the breaker trips. Using this method, the faulty zone within non-controllable switches is eventually located. Two experimental switching procedures are commonly used: the zone-by-zone rolling and the bisection method [149].

Fig. 6.3 shows the sequential search for the faulted segment for a feeder and how the adjacent feeder energizes power to the healthy segment of the feeder through the NO remote-controlled switches (RCSs). The process helps to pinpoint the fault zone. All the RCSs, along with the feeders that border the remote-controlled zones, are opened from the SCADA system after the breaker is tripped and lock out. Using the zone-by-zone rolling method, the operator proceeds by energizing the zones one by one. This starts from the feeder head (connected distribution substation) sequentially and each RCS is closed until the feeding circuit breaker trips. Once the breaker is tripped again, it is obvious that the fault zone is at the immediate RCS and that switch must remain open along with the other immediate downstream as well. Once the two RCSs are open, the breaker at the feeder head is closed and so is the NO from another feeder. This step will distinguish between RCSs and will minimize outage during the inference process from the remote control center. This process can reduce the time of search and isolate the fault based on the topological statuses updated from the SCADA system. However, there may be many non-remote-controlled switches between boundary RCSs. The crew can, however, go along those de-energized boundary RCSs to trace and pinpoint the exact fault location. As this sequential search from the feeder head can be slower (trial and error), combinations of the bisection method and then the zone-by-zone rolling may expedite the search time.

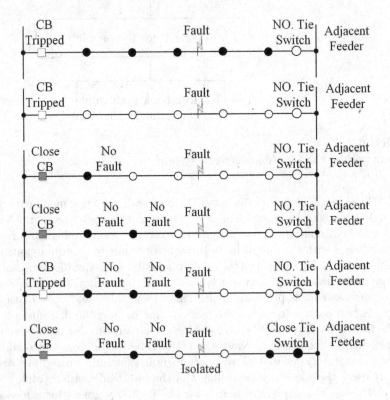

FIGURE 6.3
The zone-by-zone rolling method in simplified network topology with only remote-controlled switches.

Fig. 6.4 illustrates the search for a faulted segment using the bisection method. Among a feeder, the operator is recommended to pick the RCS located somewhere around the middle of the unsupplied feeder according to the suggestion generated from the fault localization application. The referred "half" is the middle RCS to be opened. Thereafter, the closed status of the breaker is then sent by the operator from the control center.

Prior to selecting an open switch, all relevant RCSs are first opened. The circuit breaker is then closed and if the breaker stays closed, the faulted zone is concluded such that the fault is located downstream. The second closing of RCS in the fourth row confirms that the fault is not upstream of the first open RCS. This ensures that the fault can be between the NO switch or one of the other two zones. Similar attempts have been to open sequentially one by one from the NO switch. Finally, it is confirmed the fault segment that is in the middle of the two RCSs. Meanwhile, the healthy part of the areas is gradually energized by going through a zone-by-zone rolling process.

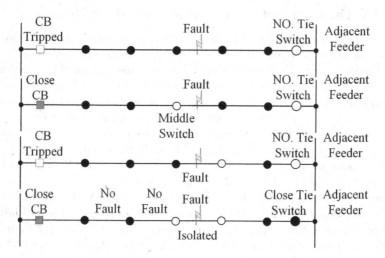

FIGURE 6.4
The bisection method.

In comparing of the two cases, the search result can be different depending on the location. The bisection method is faster than zone-by-zone rolling; however, the customers may experience multiple temporary outages. This may be safety-related because the short-circuit current might flow through the network multiple times. The lifecycle of the switches can be affected by this "trial-and-error" process [149]. The conclusion is that fault area detection algorithms may vary. The results are related to the number of deployed RCSs in a feeder and the availability of neighborhood NO switches to energize the other part of outaged feeder.

Once the fault area is identified and isolated by operators from the control center, the search for the exact location starts. If a fault location cannot be efficiently allocated by tracing along the faulted feeder, a similar process of trial-and-error may be necessary, but the crew members will have to coordinate with the operators in the control center in order to perform a narrow search of an area between non-remote-controlled switches. The bisection method is largely used in practice because zone-by-zone rolling can be too slow if there are tens of areas in a feeder. After the smallest fault area has been identified and isolated, the field crew begins the repairs on the distribution line segment as soon as possible. The restoration of power supply for the rest of the customers takes place upon completion of repair. Finally, the operator resumes the relevant RCSs back to the normal (original) switching states. The entire process of restoration may take hours, but it can significantly reduce the traditional way to exhaustively search for the fault, which may sometimes take days [149].

6.2.2　LV Faults (Faults Occur on a Secondary Network)

The LV-network has a more limited geographical area and a relatively lower level of automation. Although most of the operational steps illustrated in Fig. 6.2 are also available in LV-fault management, they are controlled by different operators. Generally, there are no remote-controlled switches (RCS) with fault indication functions in the LV-network, so that fault management was almost triggered by a trouble call from a customer experiencing a power outage or appliance failure. With the development of automatic meter reading (AMR) devices and the application of AMI nowadays, distribution control centers can receive spontaneous alarms from "smart" devices and the LV-fault can be repaired promptly before the customer notices anything.

Today, the operators typically send remote queries to AMR and/or AMI devices in the presumably faulted LV-network to determine if an actual fault has occurred in the LV-network and/or if the fault is in the customer's appliances. The fault that occurs in the LV network can be relatively quick as the affected area is between the distribution transformer (DT) and the consumer's home(s). This may not require significant efforts from the operators for a couple reasons. First, there is no direct control to the devices in the secondary network. Second, the search area is rather small compared to a long distance distribution feeder. In some cases, the communication may be limited to the customer billing center and crew members who visit the premises of the area.

6.2.3　Additional Requirements

In addition to the fault localization, isolation, and temporary service restoration based on the techniques described above, a complete fault management system has to meet some other requirements [148] in the distribution control center (DCC).

- **Simultaneous handling of multiple faults.** One (or multiple) new fault(s) could occur inside or outside the faulted area in a large distribution system while the clearance of the existing fault is not completed. The fault management system should be able to detect and deal with several faults simultaneously.

- **Automatic trigger of localization and isolation.** Once a fault occurs in the system, the fault localization and isolation analysis should start automatically without any operator or crew interaction. The management system can gather and collect all related information about the fault (produced by the fault protection equipment or indicators in the distribution network). The faulted area isolation algorithm has to be started automatically after the end of the fault localization. The generated diagnosis and recommendations for remedial actions should be sent to operators or managers once the analysis is completed.

- **Switching strategies for isolation or restoration.** To update the diagnosis and recommendations for remedial actions after receiving the latest feedback (e.g., information from the field crew), the isolation and temporary restoration analysis must be capable of being restarted by operators under some necessary situations.

- **Constraints check for system restoration.** For each combination of remote-controlled switches (RCS) proposed in the analysis of the system restoration, a power flow analysis should be executed, assuming the candidate switch sets are closed. The loads in the power-restored area can be assumed with their pre-fault value. The result of this power flow is checked for constraint violation. If there is no limit violation, this set of switches can be entered in the list of recommended operations for system restoration. Switch combinations that will cause the constraint violations (such as overload) will be removed from the restoration switch list.

- **Management system of switching procedure.** Once the proposal of isolation and restoration is approved, the corresponding remedial switching actions will be performed. The created switching procedure plan should be sent to the switch management system. The relative actions then can be executed automatically by this management function.

- **Recording of all remedial actions.** All proposed remedial switching procedures approved by the operator or manager have to be recorded by the fault management system. This is necessary to support the operator for future fault clearance when (s)he has to reconfigure the network to the pre-fault status.

- **Generation of outage reports.** The outage report for the fault should be archived for future usage. This report should contain:

    - Time of occurrence;
    - Fault location;
    - De-energized equipment;

- Number of de-energized loads;
- Number of power outage customers;
- Unsupplied power (kW or kVA) and energy usage (kWh);
- Selected switch sets for isolation;
- Selected switch sets for restoration;
- Outage duration for customers;
- Maximum outage duration;
- Filed crew reports;
- Operator or manager comments.

All the requirements listed above have to be combined with an intuitive and user-friendly human-system interface.

6.3   Graph-Based Fault Isolation

The goal of fault isolation is to separate the faulted area of the network from the faultless parts. The common approach is to assume that all switches are open in the de-energized area, and that at least one terminal belongs to the faultless region within the de-energized area. This operation is made twice, first using only remote-controlled switches and second using remote-controlled and uncontrollable switches together. Based on each resulting topology, the network groups are recalculated. Now all switches formerly simulated as open are checked to see if they are now separating the faulted region from the fault-free section. If so, this switch is included into the isolation proposal. The results of these calculations are two lists of isolation proposals: The first contains a list of remote-controlled switches, which would isolate the faulted segment from the rest of the network. The second list contains all non-remote-controlled switches, which can be opened additionally to improve the isolation. This means the amount of faultless equipment which is still connected to the faulted section is reduced. The separation should ensure that the number of faultless equipment connected to the faulted segment is minimized [147, 148, 150].

In the graph representation of a distribution network, after locating the faulted area, the border remote-controlled switches connected to this area will be opened to isolate the faulted segment from the faultless section. Use the topology of a feeder shown in Fig. 6.5 as an example to demonstrate the graph-based fault isolation procedure. Under normal conditions, the feeder is energized from the breaker. This topology contains 12 nodes and 5 switches. The lines and switches are topologically connected. Breaker and recloser are equipped with the relay. The remainder of other switching devices is the pole-mounted controlled switches that come with a fault indicator, shown as FI in the diagram for each device.

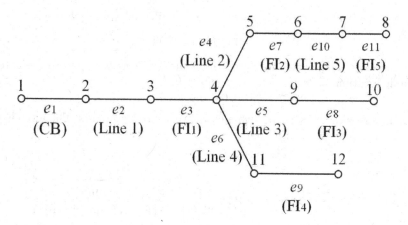

FIGURE 6.5
A radial topology with 12 nodes and 11 edges.

6.3.1 Scenario 1: Single Yes and Multiple No FIs

In this case, assume FI_1 is "Yes" while FI_2, FI_3, FI_4, and FI_5 are shown as "No."

Step 1

Find the fault segment according to Chapter 5. Three initializations are the topological incidence matrix:

$$M_i = \begin{array}{c} \\ v_1 \\ v_2 \\ v_3 \\ v_4 \\ v_5 \\ v_6 \\ v_7 \\ v_8 \\ v_9 \\ v_{10} \\ v_{11} \\ v_{12} \end{array} \begin{array}{cccccccccccc} e_1 & e_2 & e_3 & e_4 & e_5 & e_6 & e_7 & e_8 & e_9 & e_{10} & e_{11} \\ 1 & 0 & 0 & 0 & 0 & 0 & 0 & 0 & 0 & 0 & 0 \\ 1 & 1 & 0 & 0 & 0 & 0 & 0 & 0 & 0 & 0 & 0 \\ 0 & 1 & 1 & 0 & 0 & 0 & 0 & 0 & 0 & 0 & 0 \\ 0 & 0 & 1 & 1 & 1 & 1 & 0 & 0 & 0 & 0 & 0 \\ 0 & 0 & 0 & 1 & 0 & 0 & 1 & 0 & 0 & 0 & 0 \\ 0 & 0 & 0 & 0 & 0 & 0 & 1 & 0 & 0 & 1 & 0 \\ 0 & 0 & 0 & 0 & 0 & 0 & 0 & 0 & 0 & 1 & 1 \\ 0 & 0 & 0 & 0 & 0 & 0 & 0 & 0 & 0 & 0 & 1 \\ 0 & 0 & 0 & 0 & 1 & 0 & 0 & 1 & 0 & 0 & 0 \\ 0 & 0 & 0 & 0 & 0 & 0 & 0 & 1 & 0 & 0 & 0 \\ 0 & 0 & 0 & 0 & 0 & 1 & 0 & 0 & 1 & 0 & 0 \\ 0 & 0 & 0 & 0 & 0 & 0 & 0 & 0 & 1 & 0 & 0 \end{array} ,$$

the edge vector is

$$E_{\overline{FI}} = [1\ 1\ 0\ 1\ 1\ 1\ 1\ 1\ 1\ 1\ 1],$$

and the initial vertex indication vector for the affected area is shown as:

$$V_D = [1\ 0\ 0\ 0\ 0\ 0\ 0\ 0\ 0\ 0\ 0\ 0].$$

After the fault localization analysis, the generated vertex vector to indicate the fault location:

$$E_D^{\text{fault}} = [1\ 1\ 0\ 0\ 0\ 0\ 1\ 1\ 1\ 1\ 1],$$

which indicates the edges 3, 4, 5, and 6 are the faulted segments. The intended isolated area is shown in Fig. 6.6.

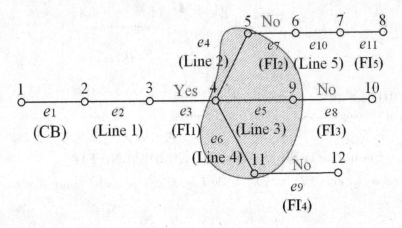

FIGURE 6.6
The intended isolated area in the radial topology for scenario 1.

Step 2

Compare the index set of all remote-controlled switches with fault indicators

$$E_D^{\text{switch}} = [3\ 7\ 8\ 9\ 11]$$

with the index set of all affected segments produced from the previous step

$$E_D^{\text{Aff}} = [3\ 4\ 5\ 6].$$

The coincident element between these two sets is 3, which indicates e_3, and it is one of the remote-controlled switches in this case.

Step 3

Remove the element(s) generated in the previous step from the affected segments set to get a new edge index set:

$$E_D^{\text{Aff,Seg}} = [4\ 5\ 6],$$

in which the corresponding edges are all line segments. Traverse all elements between columns 4, 5, and 6 in the incidence matrix to find out the latter non-zero element on each column (assume the network is topologically reordered), as shown in the incidence matrix, to record their indexes.

$$
M_i = \begin{array}{c} \\ v_1 \\ v_2 \\ v_3 \\ v_4 \\ v_5 \\ v_6 \\ v_7 \\ v_8 \\ v_9 \\ v_{10} \\ v_{11} \\ v_{12} \end{array}
\begin{array}{ccccccccccc}
e_1 & e_2 & e_3 & e_4 & e_5 & e_6 & e_7 & e_8 & e_9 & e_{10} & e_{11} \\
1 & 0 & 0 & 0 & 0 & 0 & 0 & 0 & 0 & 0 & 0 \\
1 & 1 & 0 & 0 & 0 & 0 & 0 & 0 & 0 & 0 & 0 \\
0 & 1 & 1 & 0 & 0 & 0 & 0 & 0 & 0 & 0 & 0 \\
0 & 0 & 1 & 1 & 1 & 1 & 0 & 0 & 0 & 0 & 0 \\
0 & 0 & 0 & 1 & 0 & 0 & 1 & 0 & 0 & 0 & 0 \\
0 & 0 & 0 & 0 & 0 & 0 & 1 & 0 & 0 & 1 & 0 \\
0 & 0 & 0 & 0 & 0 & 0 & 0 & 0 & 0 & 1 & 1 \\
0 & 0 & 0 & 0 & 0 & 0 & 0 & 0 & 0 & 0 & 1 \\
0 & 0 & 0 & 0 & 1 & 0 & 0 & 1 & 0 & 0 & 0 \\
0 & 0 & 0 & 0 & 0 & 0 & 0 & 1 & 0 & 0 & 0 \\
0 & 0 & 0 & 0 & 0 & 1 & 0 & 0 & 1 & 0 & 0 \\
0 & 0 & 0 & 0 & 0 & 0 & 0 & 0 & 1 & 0 & 0
\end{array}.
$$

Then, check the rows where the non-zero elements are located to find the other non-zero elements and extract their indexes.

$$
M_i = \begin{array}{c} \\ v_1 \\ v_2 \\ v_3 \\ v_4 \\ v_5 \\ v_6 \\ v_7 \\ v_8 \\ v_9 \\ v_{10} \\ v_{11} \\ v_{12} \end{array}
\begin{array}{ccccccccccc}
e_1 & e_2 & e_3 & e_4 & e_5 & e_6 & e_7 & e_8 & e_9 & e_{10} & E_{11} \\
1 & 0 & 0 & 0 & 0 & 0 & 0 & 0 & 0 & 0 & 0 \\
1 & 1 & 0 & 0 & 0 & 0 & 0 & 0 & 0 & 0 & 0 \\
0 & 1 & 1 & 0 & 0 & 0 & 0 & 0 & 0 & 0 & 0 \\
0 & 0 & 1 & 1 & 1 & 1 & 0 & 0 & 0 & 0 & 0 \\
0 & 0 & 0 & 1 & 0 & 0 & 1 & 0 & 0 & 0 & 0 \\
0 & 0 & 0 & 0 & 0 & 0 & 1 & 0 & 0 & 1 & 0 \\
0 & 0 & 0 & 0 & 0 & 0 & 0 & 0 & 0 & 1 & 1 \\
0 & 0 & 0 & 0 & 0 & 0 & 0 & 0 & 0 & 0 & 1 \\
0 & 0 & 0 & 0 & 1 & 0 & 0 & 1 & 0 & 0 & 0 \\
0 & 0 & 0 & 0 & 0 & 0 & 0 & 1 & 0 & 0 & 0 \\
0 & 0 & 0 & 0 & 0 & 1 & 0 & 0 & 1 & 0 & 0 \\
0 & 0 & 0 & 0 & 0 & 0 & 0 & 0 & 1 & 0 & 0
\end{array}.
$$

Therefore, the candidate vector of isolated segments is

$$E_{\text{candidate}} = [4\ 7\ 5\ 8\ 6\ 9].$$

Step 4

Compare the $E_{\text{candidate}}$ with the E_D^{switch} to identify the remote-controlled switches in the index vector of isolated segments. The intersection of the sets is

$$E_{\text{temp}} = [7\ 8\ 9].$$

Then, combine the element(s) generated from Step 2 to create the complete list of boundary switches for the faulted segment as:

$$E_{\text{Bound}} = [3 \ 7 \ 8 \ 9].$$

Step 5

Set all column values to be zeros for columns $[3 \ 7 \ 8 \ 9]$ in order to generate the topology incidence matrix after the fault isolation operation.

$$M_i = \begin{array}{c} \\ v_1 \\ v_2 \\ v_3 \\ v_4 \\ v_5 \\ v_6 \\ v_7 \\ v_8 \\ v_9 \\ v_{10} \\ v_{11} \\ v_{12} \end{array} \begin{pmatrix} e_1 & e_2 & e_3 & e_4 & e_5 & e_6 & e_7 & e_8 & e_9 & e_{10} & e_{11} \\ 1 & 0 & 0 & 0 & 0 & 0 & 0 & 0 & 0 & 0 & 0 \\ 1 & 1 & 0 & 0 & 0 & 0 & 0 & 0 & 0 & 0 & 0 \\ 0 & 1 & 0 & 0 & 0 & 0 & 0 & 0 & 0 & 0 & 0 \\ 0 & 0 & 0 & 1 & 1 & 1 & 0 & 0 & 0 & 0 & 0 \\ 0 & 0 & 0 & 1 & 0 & 0 & 0 & 0 & 0 & 0 & 0 \\ 0 & 0 & 0 & 0 & 0 & 0 & 0 & 0 & 0 & 1 & 0 \\ 0 & 0 & 0 & 0 & 0 & 0 & 0 & 0 & 0 & 1 & 1 \\ 0 & 0 & 0 & 0 & 0 & 0 & 0 & 0 & 0 & 0 & 1 \\ 0 & 0 & 0 & 0 & 1 & 0 & 0 & 0 & 0 & 0 & 0 \\ 0 & 0 & 0 & 0 & 0 & 0 & 0 & 0 & 0 & 0 & 0 \\ 0 & 0 & 0 & 0 & 0 & 1 & 0 & 0 & 0 & 0 & 0 \\ 0 & 0 & 0 & 0 & 0 & 0 & 0 & 0 & 0 & 0 & 0 \end{pmatrix},$$

6.3.2   Scenario 2: Multiple Yes and Single/Multiple No FIs

In this case, assume FI_1 and FI_2 are "Yes," while FI_3, FI_4, and FI_5 are shown as "No."

Step 1

After the fault localization analysis, the generated vertex vector to indicate the fault location:

$$E_D^{\text{fault}} = [1 \ 1 \ 1 \ 1 \ 1 \ 1 \ 0 \ 1 \ 1 \ 0 \ 1]$$

which indicates the edges 7 and 10 are the faulted segments. The intended isolated area is shown in Fig. 6.7.

Step 2

Compare the index set of all remote-controlled switches with fault indicators

$$E_D^{\text{switch}} = [3 \ 7 \ 8 \ 9 \ 11]$$

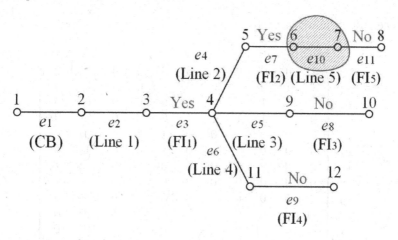

FIGURE 6.7
The intended isolated area in the radial topology for scenario 2.

with the index set of all affected segments produced from the previous step

$$E_D^{\text{Aff}} = [7\ 10].$$

The coincident element between these two sets is 7, which indicates e_7, and it is one of the remote-controlled switches in this case.

Step 3

Remove the element(s) generated in the previous step from the affected segments set and get a new edge index set

$$E_D^{\text{Aff,Seg}} = [10],$$

in which the corresponding edge is a line segment. Traverse all elements in column 10 in the incidence matrix to find out the latter non-zero element on each column (assume the network is topologically reordered) as shown in the incidence matrix, to record their indexes.

$$M_i = \begin{array}{c|ccccccccccc} & e_1 & e_2 & e_3 & e_4 & e_5 & e_6 & e_7 & e_8 & e_9 & e_{10} & e_{11} \\ v_1 & 1 & 0 & 0 & 0 & 0 & 0 & 0 & 0 & 0 & 0 & 0 \\ v_2 & 1 & 1 & 0 & 0 & 0 & 0 & 0 & 0 & 0 & 0 & 0 \\ v_3 & 0 & 1 & 1 & 0 & 0 & 0 & 0 & 0 & 0 & 0 & 0 \\ v_4 & 0 & 0 & 1 & 1 & 1 & 1 & 0 & 0 & 0 & 0 & 0 \\ v_5 & 0 & 0 & 0 & 1 & 0 & 0 & 1 & 0 & 0 & 0 & 0 \\ v_6 & 0 & 0 & 0 & 0 & 0 & 0 & 1 & 0 & 0 & 1 & 0 \\ v_7 & 0 & 0 & 0 & 0 & 0 & 0 & 0 & 0 & 0 & 1 & 1 \\ v_8 & 0 & 0 & 0 & 0 & 0 & 0 & 0 & 0 & 0 & 0 & 1 \\ v_9 & 0 & 0 & 0 & 0 & 1 & 0 & 0 & 1 & 0 & 0 & 0 \\ v_{10} & 0 & 0 & 0 & 0 & 0 & 0 & 0 & 1 & 0 & 0 & 0 \\ v_{11} & 0 & 0 & 0 & 0 & 0 & 1 & 0 & 0 & 1 & 0 & 0 \\ v_{12} & 0 & 0 & 0 & 0 & 0 & 0 & 0 & 0 & 1 & 0 & 0 \end{array}.$$

Then, check the rows where the non-zero elements are located to find the other non-zero elements, and extract their indexes.

$$M_i = \begin{array}{c|ccccccccccc} & e_1 & e_2 & e_3 & e_4 & e_5 & e_6 & e_7 & e_8 & e_9 & e_{10} & e_{11} \\ v_1 & 1 & 0 & 0 & 0 & 0 & 0 & 0 & 0 & 0 & 0 & 0 \\ v_2 & 1 & 1 & 0 & 0 & 0 & 0 & 0 & 0 & 0 & 0 & 0 \\ v_3 & 0 & 1 & 1 & 0 & 0 & 0 & 0 & 0 & 0 & 0 & 0 \\ v_4 & 0 & 0 & 1 & 1 & 1 & 1 & 0 & 0 & 0 & 0 & 0 \\ v_5 & 0 & 0 & 0 & 1 & 0 & 0 & 1 & 0 & 0 & 0 & 0 \\ v_6 & 0 & 0 & 0 & 0 & 0 & 0 & 1 & 0 & 0 & 1 & 0 \\ v_7 & 0 & 0 & 0 & 0 & 0 & 0 & 0 & 0 & 0 & 1 & 1 \\ v_8 & 0 & 0 & 0 & 0 & 0 & 0 & 0 & 0 & 0 & 0 & 1 \\ v_9 & 0 & 0 & 0 & 0 & 1 & 0 & 0 & 1 & 0 & 0 & 0 \\ v_{10} & 0 & 0 & 0 & 0 & 0 & 0 & 0 & 1 & 0 & 0 & 0 \\ v_{11} & 0 & 0 & 0 & 0 & 0 & 1 & 0 & 0 & 1 & 0 & 0 \\ v_{12} & 0 & 0 & 0 & 0 & 0 & 0 & 0 & 0 & 1 & 0 & 0 \end{array}.$$

Therefore, the candidate vector of isolated segments is

$$E_{\text{candidate}} = [10\ 11].$$

Step 4

Compare the $E_{\text{candidate}}$ with the E_D^{switch} to identify the remote-controlled switches in the index vector of isolated segments. The intersection of the sets is

$$E_{\text{temp}} = [11].$$

Then, combine the element(s) generated from Step 2 to create the complete list of boundary switches for the faulted segment:

$$E_{\text{Bound}} = [7 \; 11].$$

Step 5

Set all column values as zero for columns [7 11] to generate the topology incidence matrix after the fault isolation operation.

$$M_i = \begin{array}{c} \\ v_1 \\ v_2 \\ v_3 \\ v_4 \\ v_5 \\ v_6 \\ v_7 \\ v_8 \\ v_9 \\ v_{10} \\ v_{11} \\ v_{12} \end{array} \begin{array}{c} \begin{array}{ccccccccccc} e_1 & e_2 & e_3 & e_4 & e_5 & e_6 & e_7 & e_8 & e_9 & e_{10} & e_{11} \end{array} \\ \left(\begin{array}{ccccccccccc} 1 & 0 & 0 & 0 & 0 & 0 & 0 & 0 & 0 & 0 & 0 \\ 1 & 1 & 0 & 0 & 0 & 0 & 0 & 0 & 0 & 0 & 0 \\ 0 & 1 & 0 & 0 & 0 & 0 & 0 & 0 & 0 & 0 & 0 \\ 0 & 0 & 0 & 1 & 1 & 1 & 0 & 0 & 0 & 0 & 0 \\ 0 & 0 & 0 & 1 & 0 & 0 & 0 & 0 & 0 & 0 & 0 \\ 0 & 0 & 0 & 0 & 0 & 0 & 0 & 0 & 0 & 1 & 0 \\ 0 & 0 & 0 & 0 & 0 & 0 & 0 & 0 & 0 & 1 & 0 \\ 0 & 0 & 0 & 0 & 0 & 0 & 0 & 0 & 0 & 0 & 0 \\ 0 & 0 & 0 & 0 & 1 & 0 & 0 & 0 & 0 & 0 & 0 \\ 0 & 0 & 0 & 0 & 0 & 0 & 0 & 0 & 0 & 0 & 0 \\ 0 & 0 & 0 & 0 & 0 & 1 & 0 & 0 & 0 & 0 & 0 \\ 0 & 0 & 0 & 0 & 0 & 0 & 0 & 0 & 0 & 0 & 0 \end{array} \right) \end{array},$$

The corresponding MATLAB sample code is illustrated in Listing 6.1. It should be noted that the input variables of this program are generated from the fault localization algorithm.

Listing 6.1
Fault Isolation Analysis

```
1   clear all
2   close all
3   clc
4
5   % Fault segment localization to find fault segments from ...
        Chapter 5
6   % Input the Incidence matrix, EdgeVec(E_D), and NodeVec(V_D).
7   edge_state = floc(IncM, E_D, V_D);
8
9   sw_ind = input('input the index of switches: ');
10
11  % Index of fault segments and affected switches.
12  affect_edge1 = find(edge_state == 0);
13
14  % Open the affected switch first.
15  open_sw1 = intersect(sw_ind, affect_edge1);
16  for n = 1:length(open_sw1)
17      m =open_sw1(n);
```

```
18        edge_state(m) = 1;
19   end
20
21   % Index of fault segments only.
22   affect_edge2 = find(edge_state == 0);
23
24   % Find switches connected with fault segments.
25   for n = 1:length(affect_edge2)
26        m = affect_edge2(n);
27        affect_node = find(IncM(:,m) == 1,1,'last');
28        connect_edge(n,:) = find(IncM(affect_node,:) == 1);
29   end
30   connect_edge = reshape(connect_edge,1,numel(connect_edge));
31   open_sw2 = intersect(sw_ind,connect_edge);
32   open_sw = [open_sw1,open_sw2];
33
34   % Break all switches connected with fault segments in ...
               incidence matrix.
35   for i = 1:length(open_sw)
36        j = open_sw(i);
37        IncM(:,j) = zeros(size(IncM,1),1);
38   end
```

6.4　Temporary Service Restoration

Once the operator successfully isolates the faulted area, the temporary restoration to the healthy part downstream of the feeder can only happen if there are one or more tie RCSs connected to the other feeders that have multiple options for reconfiguration. If a feeder is not connected to any NO switch, the outage time can be longer and the number of affected customers can be significant. This can degrade overall system reliability. The healthy part will still be energized in the radial topology but from another feeder [151, 152].

Since the restoration plan will be provided to operators in the management center to help them make a decision on how to restore the power outage area temporarily without power supply barriers and damage to the whole system, it must meet the practical requirements of system dispatchers as summarized below [153–156]:

- Estimate ample outage time and anticipation of restoration time as there will often be more complaints from customers if it is otherwise. Automation of remote switching actually helps significantly to pinpoint potential fault areas;

- Restore as much load as possible within the power outage area but such that no components are overloaded;

- Radial system topology must be retained in any possible reconfiguration;

- The required number of switching operations in the restoration plan should be minimal.

Switching frequency should be planned out minimally in order to avoid interruption. At the same time, the least switching may reduce the life expectancy of switches. In addition, it takes time for operators or field crews to operate a switch (remote or non-remote controllable). This is an especially crucial factor in a crowded city because it may take the field crew more than one hour to get to a place only one mile away;

- The new reconfigured network topology that is temporarily arranged should be closest to the normal state. This can minimize interruption of services when a repair of a line is complete.

As shown in Fig. 6.8, this is a 2-feeder example with 10 nodes and 9 edges. The nodes 1 and 10 are two distribution substations. Edge 5 is a normally open tie switch while the other edges are remote-controlled, normally closed (NC) switches. The nodes and edges are electrically connected. For the intuitiveness and convenience analysis, the detailed distribution elements of alternative feeders, or substations between two switches are deleted. The demonstration of candidate substations connected to a line segment between two remote-controlled switches is illustrated in Fig. 6.9. All of these backup feeders and substations can improve the reconfigurability of the whole system. The corresponding incidence matrix of Fig. 6.8 is shown as:

$$
M_i =
\begin{array}{c}
\\ v_1 \\ v_2 \\ v_3 \\ v_4 \\ v_5 \\ v_6 \\ v_7 \\ v_8 \\ v_9 \\ v_{10}
\end{array}
\begin{array}{c}
\begin{array}{ccccccccc}
e_1 & e_2 & e_3 & e_4 & e_5 & e_6 & e_7 & e_8 & e_9
\end{array} \\
\left(
\begin{array}{ccccccccc}
1 & 0 & 0 & 0 & 0 & 0 & 0 & 0 & 0 \\
1 & 1 & 0 & 0 & 0 & 0 & 0 & 0 & 0 \\
0 & 1 & 1 & 0 & 0 & 0 & 0 & 0 & 0 \\
0 & 0 & 1 & 1 & 0 & 0 & 0 & 0 & 0 \\
0 & 0 & 0 & 1 & 0 & 0 & 0 & 0 & 0 \\
0 & 0 & 0 & 0 & 0 & 1 & 0 & 0 & 0 \\
0 & 0 & 0 & 0 & 0 & 1 & 1 & 0 & 0 \\
0 & 0 & 0 & 0 & 0 & 0 & 1 & 1 & 0 \\
0 & 0 & 0 & 0 & 0 & 0 & 0 & 1 & 1 \\
0 & 0 & 0 & 0 & 0 & 0 & 0 & 0 & 1
\end{array}
\right)
\end{array} .
$$

If a fault occurs between remote-controlled switches 2 and 3, the incidence

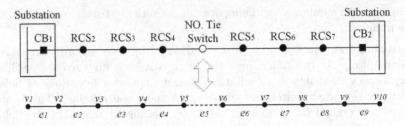

FIGURE 6.8
A 2-feeder example with its graph representation.

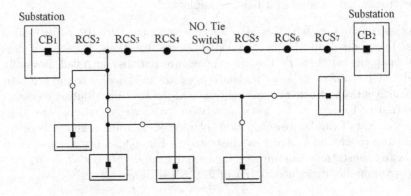

FIGURE 6.9
An example of the interconnected feeders with multiple NO switches.

matrix generated after the fault isolation operation will be:

$$
M_i =
\begin{array}{c}
\\
v_1 \\
v_2 \\
v_3 \\
v_4 \\
v_5 \\
v_6 \\
v_7 \\
v_8 \\
v_9 \\
v_{10}
\end{array}
\begin{array}{c}
\begin{array}{cccccccccc}
e_1 & e_2 & e_3 & e_4 & e_5 & e_6 & e_7 & e_8 & e_9
\end{array} \\
\left(
\begin{array}{ccccccccc}
1 & 0 & 0 & 0 & 0 & 0 & 0 & 0 & 0 \\
1 & 0 & 0 & 0 & 0 & 0 & 0 & 0 & 0 \\
0 & 0 & 0 & 0 & 0 & 0 & 0 & 0 & 0 \\
0 & 0 & 0 & 1 & 0 & 0 & 0 & 0 & 0 \\
0 & 0 & 0 & 1 & 0 & 0 & 0 & 0 & 0 \\
0 & 0 & 0 & 0 & 0 & 1 & 0 & 0 & 0 \\
0 & 0 & 0 & 0 & 0 & 1 & 1 & 0 & 0 \\
0 & 0 & 0 & 0 & 0 & 0 & 1 & 1 & 0 \\
0 & 0 & 0 & 0 & 0 & 0 & 0 & 1 & 1 \\
0 & 0 & 0 & 0 & 0 & 0 & 0 & 0 & 1
\end{array}
\right)
\end{array}.
$$

Technically, the operator will close the normally open tie switch e_5 to provide the temporary power supply for the "healthy" area (between remote-controlled switches 3 and 5). Then, he or she will find the corresponding column of e_5 (all zeros) in the incidence matrix and connect the edge with nodes 5 and 6 as:

$$M_i = \begin{array}{c} \\ v_1 \\ v_2 \\ v_3 \\ v_4 \\ v_5 \\ v_6 \\ v_7 \\ v_8 \\ v_9 \\ v_{10} \end{array} \begin{pmatrix} \begin{array}{ccccccccc} e_1 & e_2 & e_3 & e_4 & e_5 & e_6 & e_7 & e_8 & e_9 \\ 1 & 0 & 0 & 0 & 0 & 0 & 0 & 0 & 0 \\ 1 & 0 & 0 & 0 & 0 & 0 & 0 & 0 & 0 \\ 0 & 0 & 0 & 0 & 0 & 0 & 0 & 0 & 0 \\ 0 & 0 & 0 & 1 & 0 & 0 & 0 & 0 & 0 \\ 0 & 0 & 0 & 1 & 1 & 0 & 0 & 0 & 0 \\ 0 & 0 & 0 & 0 & 1 & 1 & 0 & 0 & 0 \\ 0 & 0 & 0 & 0 & 0 & 1 & 1 & 0 & 0 \\ 0 & 0 & 0 & 0 & 0 & 0 & 1 & 1 & 0 \\ 0 & 0 & 0 & 0 & 0 & 0 & 0 & 1 & 1 \\ 0 & 0 & 0 & 0 & 0 & 0 & 0 & 0 & 1 \end{array} \end{pmatrix}.$$

Alternatively, the adjacency matrix of the 2-feeder example:

$$M_a = \begin{array}{c} \\ v_1 \\ v_2 \\ v_3 \\ v_4 \\ v_5 \\ v_6 \\ v_7 \\ v_8 \\ v_9 \\ v_{10} \end{array} \begin{pmatrix} \begin{array}{cccccccccc} v_1 & v_2 & v_3 & v_4 & v_5 & v_6 & v_7 & v_8 & v_9 & v_{10} \\ 0 & 1 & 0 & 0 & 0 & 0 & 0 & 0 & 0 & 0 \\ 1 & 0 & 1 & 0 & 0 & 0 & 0 & 0 & 0 & 0 \\ 0 & 1 & 0 & 1 & 0 & 0 & 0 & 0 & 0 & 0 \\ 0 & 0 & 1 & 0 & 1 & 0 & 0 & 0 & 0 & 0 \\ 0 & 0 & 0 & 1 & 0 & 0 & 0 & 0 & 0 & 0 \\ 0 & 0 & 0 & 0 & 0 & 0 & 1 & 0 & 0 & 0 \\ 0 & 0 & 0 & 0 & 0 & 1 & 0 & 1 & 0 & 0 \\ 0 & 0 & 0 & 0 & 0 & 0 & 1 & 0 & 1 & 0 \\ 0 & 0 & 0 & 0 & 0 & 0 & 0 & 1 & 0 & 1 \\ 0 & 0 & 0 & 0 & 0 & 0 & 0 & 0 & 1 & 0 \end{array} \end{pmatrix},$$

and the fault isolated form:

$$M_a = \begin{array}{c} \\ v_1 \\ v_2 \\ v_3 \\ v_4 \\ v_5 \\ v_6 \\ v_7 \\ v_8 \\ v_9 \\ v_{10} \end{array} \begin{pmatrix} \begin{array}{cccccccccc} v_1 & v_2 & v_3 & v_4 & v_5 & v_6 & v_7 & v_8 & v_9 & v_{10} \\ 0 & 1 & 0 & 0 & 0 & 0 & 0 & 0 & 0 & 0 \\ 1 & 0 & 0 & 0 & 0 & 0 & 0 & 0 & 0 & 0 \\ 0 & 0 & 0 & 0 & 0 & 0 & 0 & 0 & 0 & 0 \\ 0 & 0 & 0 & 0 & 1 & 0 & 0 & 0 & 0 & 0 \\ 0 & 0 & 0 & 1 & 0 & 0 & 0 & 0 & 0 & 0 \\ 0 & 0 & 0 & 0 & 0 & 0 & 1 & 0 & 0 & 0 \\ 0 & 0 & 0 & 0 & 0 & 1 & 0 & 1 & 0 & 0 \\ 0 & 0 & 0 & 0 & 0 & 0 & 1 & 0 & 1 & 0 \\ 0 & 0 & 0 & 0 & 0 & 0 & 0 & 1 & 0 & 1 \\ 0 & 0 & 0 & 0 & 0 & 0 & 0 & 0 & 1 & 0 \end{array} \end{pmatrix}.$$

Similarly, the temporary service restoration operation is to connect the nodes

5 and 6 in the matrix form as:

$$
M_a =
\begin{array}{c@{}c}
 & \begin{array}{cccccccccc} v_1 & v_2 & v_3 & v_4 & v_5 & v_6 & v_7 & v_8 & v_9 & v_{10} \end{array} \\
\begin{array}{c} v_1 \\ v_2 \\ v_3 \\ v_4 \\ v_5 \\ v_6 \\ v_7 \\ v_8 \\ v_9 \\ v_{10} \end{array} &
\left(\begin{array}{cccccccccc}
0 & 1 & 0 & 0 & 0 & 0 & 0 & 0 & 0 & 0 \\
1 & 0 & 0 & 0 & 0 & 0 & 0 & 0 & 0 & 0 \\
0 & 0 & 0 & 0 & 0 & 0 & 0 & 0 & 0 & 0 \\
0 & 0 & 0 & 0 & 1 & 0 & 0 & 0 & 0 & 0 \\
0 & 0 & 0 & 1 & 0 & 1 & 0 & 0 & 0 & 0 \\
0 & 0 & 0 & 0 & 1 & 0 & 1 & 0 & 0 & 0 \\
0 & 0 & 0 & 0 & 0 & 1 & 0 & 1 & 0 & 0 \\
0 & 0 & 0 & 0 & 0 & 0 & 1 & 0 & 1 & 0 \\
0 & 0 & 0 & 0 & 0 & 0 & 0 & 1 & 0 & 1 \\
0 & 0 & 0 & 0 & 0 & 0 & 0 & 0 & 1 & 0
\end{array}\right)
\end{array}.
$$

It should be noted that the sequential number ordering of edges and nodes should be quite organized. Without that, it is hard to tell the topological feasibility and the corresponding operations only according to the matrix form. In the 2-feeder example, the upper-right and lower-left parts in the matrix forms (incidence and adjacency) are all zeros, the service restoration operation processed between the other two parts of the containing elements. If the ordering changes determining NO switches by inspection is not made obvious but is still feasible.

The corresponding MATLAB®sample code is illustrated in Listing 6.2. It should be noted that the input variables of this program are generated from the fault isolation algorithm in Listing 6.1.

Listing 6.2
Temporary service restoration

```
1  clear all
2  close all
3  clc
4
5  % Initialization: input topological incidence matrix;
6  % Constructe graph based on conectivity of vertices;
7  % Input the incidence matrix generated from fault isolation;
8  re = dfsinc(IncM);
9  s=re(:,1);
10 t=re(:,2);
11 G = graph(s,t);
12 source_n = input('Input the index of cadidate source node: ');
13
14 % Depth-first-search to decide which tie switch to close.
15 for i = 1:length(source_n)
16     j = source_n(i);
17     v = dfsearch(G,j);
18     depth(i) = find(v==1);
19 end
20
21 tiesw = input('Input the index of tie switches: ');
```

```
22
23   % Find the tail node of cadidate tie switches.
24   for i = 1:length(tiesw)
25       j = tiesw(i);
26       tiesw_tail(i) = find(IncM(:,j) == 1,1,'last');
27   end
28
29
30   close_tiesw_tail = tiesw_tail(find(min(depth)));
31   close_tiesw = find(IncM(close_tiesw_tail,:) == 1,1,'first');
32   open_tiesw1 = input('Input the index of tie switch for fault ...
         isolation: ');
33   open_tiesw2 = source_n - close_tiesw;
```

6.5   Conclusions

Service restoration is crucial to overall system reliability in terms of time, frequency, and the number of customers affected by the outage. Hence, a temporary arrangement of partial restoration is important. This chapter explores the possibilities to connect normally open (NO) switches of a given feeder. The tie switches are established for this proposal and are often interfaced with power flow modules to determine the estimated power losses and voltage violation for each hypothetical scenario to be enumerated based on the other injection sources and their direct or indirect connections to the outage area.

A heuristic search with specific knowledge of experience can expedite the search. The modeling of SCADA FIs and real-time topology statuses do provide clues to the system operators. The next few chapters will relate to the additional input information such as trouble call tickets and how they could strengthen the search of fault location by coordinating with the crew at the site. Scheduled outages may also require careful evaluation of switching to de-energize a segment, and may be safety-related.

Mini Project 5: Integrate with Power Flow Module

The MATLAB scripts provided in this chapter are the modules on handling isolation of fault segments and providing partial restoration to the healthy part of a feeder. The topology networks in the case studies have multiple reconfigurations possible made other NO tie switches. Implement the power flow module (from Chapter 4) to evaluate the total losses as well as the voltage profile and potential violations with each combination. This combinatorial evaluation for each case can provide a recommendation to the dispatchers with regards to what choice would be optimal in operation. The modules in this chapter only consider remote-controlled switches (RCSs), i.e., the disconnector or recloser that is equipped with feeder remote-terminal units (FRTUs).

Part IV

Outage Coordination and Correlation

7

Customer Relational Database

The way electricity billing information has been collected over the past decades has been labor intensive. Utilities regularly have to dispatch their crews to visit customer premises and record energy usage displayed on electromechanical analog meters. The frequency of getting such data is typically either once or twice each month. While information communication technology has evolved, the ubiquitous sensor network has improved the data collection of billing information. This is typically referred to as automatic meter reading (AMR), as it was first introduced early this century, or AMI which was introduced in recent years. The challenges now are not just the large number of deployments to the sites, but also the maintenance of those IP-based "smart" energy meters that can be cost ineffective. Although some believe that these new cyber infrastructure additions help utilities to better manage their billing information, as a result of improving the accuracy of energy consumption, approximately 60% of customers nationwide (per the Federal Energy Regulatory Commission) are still going through the conventional means of data collection.

7.1   Introduction to AMI

In recent years, the smart grid vision has been widely promoted to address energy sustainability, with integration of advanced sensors and technologies into the existing power infrastructure. The current status of the power grid has been improved, with new regulations of energy policies that will enhance reliability. Through the rapid expansion of intelligent communication infrastructure, the vision varies for each country. For example, the authorities in China have set ambitious goals to clearly achieve their top-down priorities to massively deploy phasor measurement units in substation networks, whereas North America has adopted bottom-up approaches, i.e., to integrate renewable energy in transmission systems and at the same time deploy "smart" IP-based meters in distribution networks. Overall, the vision has set higher standards for electricity safety, energy usage efficiency, and environmental protection, as well as operational resiliency. AMI is one of these visions and has been prioritized for providing consumers with additional options for energy use at home.

The interdisciplinary efforts with communication experts would revolutionize the power industry with a "smarter" grid as the ultimate goal.

One benefit of being part of the AMI initiative is to enhance communication between customers and utilities. This transition provides an option for consumers to decide when they would work on their laundry, when they will start charging their electric car, when they should turn off the lights, etc. Recently, modern science and technology have enhanced the transition with the use of communication and information technologies, the promotion of environment, protection procedures, and the upgrade of operating systems; these technological advances would tremendously improve AMI communication architecture.

Other issues, including asset management and energy conservation, as well as emission reduction, largely by deploying AMI systems, would be part of the important milestones for this vision [157]. AMI is an integral part of the modern communication infrastructure for distribution grids that will be connected with the dispatching control centers.

7.1.1   Hierarchical Structure of AMI Networks

AMI is a network management system consisted of "smart" IP-based meters installed on the side of users, the data management console located in the data monitoring center, and the communication network to transmit the relevant electricity information. Modern communication technology has been extended to the home area network (HAN), local area network (LAN), and wide area network (WAN), to enhance transmission efficiency in different regions with various area sizes.

Fig. 7.1 demonstrates the fundamental structure of an AMI. The "smart" meter has the ability to measure approximate real-time electricity information, e.g., the three-phase voltage and current, real and apparent power, frequency, and energy consumption. The collected data will transmit to the data management center via the pre-setting communication network. The data transmission network can be public networks or appropriative networks such as power line communications (PLC), fixed radio frequency (RF), and the power supply administration [158]. The relevant data collected from the meters are received and stored in the data centers, which have to pre-set enough storage space, and then the gathered data will be sent to the management console for modeling and analysis. The data management system is capable of monitoring and supervising relevant electricity information via web browsers. In addition, the functions of AMI control and operation, dynamic electricity price billing, and feedback on customer service can be achieved in this system. In order to improve the security level in the management port, there should always be a firewall between data reception and the management system to ensure that only users with special access have the ability to check and administrate relevant data. The bi-directional flow of information between the consumers and the customer billing center provides data resolution up to every 15 minutes per cycle [158].

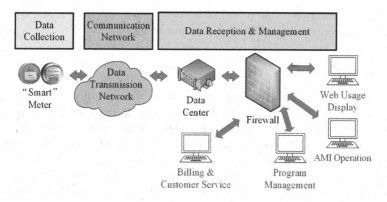

FIGURE 7.1
Communication infrastructure of an AMI system.

The AMI system is part of the communication framework for the distribution grid. Initially, the automatic meter reading (AMR) is implemented to improve the metering accuracy as well as cost reduction of labor that is constantly needed to read the kWh information at the consumers' premises. As the relative communication technology advances and the interactivity between utilities and consumers increases, the advantages and benefits of the AMI system have evolved from AMR to AMI with additional control variables on demand response of power usage. Below are the major benefits summarized for AMI:

1) Utilization of AMI information between utilities and consumers for optimal energy usage.

2) AMI provides close-to-real-time information to utilities.

3) AMI also has demand response features that allow consumers to opt in during the peak hours of the day/season.

4) AMI is a paradigm for sharing electricity usage information between consumers and the dispatching control center (DCC) when the pricing scheme would affect customers' energy usage behavior [159].

5) The detection module for power quality can be set up in "smart" meters to facilitate inspection within a distribution system [160].

6) AMI would increase the system observability that can be used in operational mode, and other applications such as cross-domain data validation against cybertampering.

7) The increased system observability would help system dispatchers to pin-point faulted segments of the secondary distribution network [161].

8) Consumers would be able to make an economical decision based on the pricing information and their desire to use energy.

7.1.2   AMR versus AMI

The major distinction between AMI and AMR is the control variables. While AMR would transmit energy usage information to the customer billing center, the AMI would enable the control capability between the utilities and customers if they choose to opt out from the peak load period that would be incentivized. The new paradigm would enable consumers to be more pro-active in participating in the electricity market. Both AMR and AMI replace the site manual reading by utility crews. The features of AMR are similar to AMI, which transfers the data of household energy consumption, meter status, as well as diagnostics to customer billing centers, mainly for collecting billing information [162]. AMI is a predecessor of AMR that provides next-generation functionalities with IP-based metering solutions. Table 7.1 shows the difference between AMI and AMR. From the users, and operational perspective, this table summarizes the important aspects of the system-wide metering implementation for a dynamic pricing market that would engage consumer participation. This IP-based metering solution would also provide information to the consumers in order for them to make a sound, economical decision. As this is not internationally used for operational purposes, some approximation of instantenous values of the power consumption/energy usage would be sufficient for the collection of household billing information.

Comparing between the two, the computational power has exponentially improved over the last decade. The embedded system of the AMI devices can be designed with a more robust capability for communication. As the power of such devices improves, including more information as well as increased frequency of information exchange is desired. In addition to that, the functionalities of the IP-based meters also include load control, prediction of potential fault location within the secondary network, and event reports such as unavailability of the meters, as well as firmware updates for enhancing the functionalities of the IP-based meters. Most upgrades are being updated with patches for data security and reliability.

7.1.3   AMI Communication Infrastructure

The communication infrastructure of an AMI consists of the IP-based devices of "smart" meters, the communication network between the customer billing center and the consumers, and the servers of meter data management systems (MDMS). These integral parts of an AMI system are described in the next subsections.

TABLE 7.1: Comparison between AMI and AMR

	AMI	**AMR**
Data Collection	Collect data according to pre-set. Support real-time reading.	Typically gather data monthly, daily at most.
Communication Mode	Two-way Communication.	One-way Communication.
Benefited Parties	Engineering, operations asset-management, planning departments, customer service, billing, and metering.	Billing and metering.
User Operation	Users can manage the working time of devices and communicate with "smart" devices.	Not applicable.
Pricing Model	Dynamic prices.	Fixed prices.

7.1.3.1 "Smart" Meter

The IP-based "smart" energy meters are the instrumentation for transmitting energy usage between customers and billing centers. They provide the data transmission interface to connect the communication network and the data acquisition unit. Fig. 7.2 shows an example of an electromechanical analog meter and an IP-based energy meter. Unlike the electromechanical analog meter, the electronic meter displays digital numbers on its panel. Crews who visit the consumer premise site would understand how to interpret the angle of arrows on the display of the electromechanical analog meter.

The measurement variables of the IP-based meters include kilowatt hour (kWh), kilowatt (KW), voltage (V), or ampere (A). It is possible to pre-set the measuring interval for 10 or 15 minutes, and then utilize the open two-way communication for the purposes of monitoring, information verification,

A.   B

FIGURE 7.2
Existing analog meter (A) and the new IP-based energy meter (B).

and diagnostics. These meters may have the privilege of receiving real-time dynamic electricity prices from the market authority, which would include demand response features.

7.1.3.2 Setup of AMI Communication

As the only medium during information transfer, the AMI communication network provides secure and effective services to ensure data exchange. Two communication modes, wired or wireless, are available in the AMI communication system. Regarding the wired mode, three main methods, which are serial communication, ethernet, and optical fiber communication, are described briefly below.

1) *Serial Communication*: This communication mode was originally designed to transfer data over large distances. In the AMI system, the IP-based meter sends data one bit at a time, sequentially, over one data cable connected with the serial communication port in the meter. Sometimes, the serial communication can also be utilized with other modes, e.g., ethernet or optical fiber [163].

2) *Ethernet Communication*: As one of the most economical and widely used modes in networking technologies for local area (LAN) or larger networks, ethernet can be used in the IP-based meters. The data cable connects the ethernet interface in the meter with the data collector for wired communication. Optimizing data transfer across multiple sites, some network protocols are available at the ethernet level, including transmission control protocol (TCP) or ModBus. ModBus can accelerate data transfer by integrating revelent electricity information into the data module installed in the IP-based meters.

3) *Optical Fiber Communication*: This mode is optimized technology from

the ethernet. This method performs rectilinear data transfer via converting electronic signals to light signals, which could send data packages to destinations with different distances immediately. The major disadvantage of this method is the exorbitant price. Generally, an AMI system collects data from a region first and then transmits via optical fiber.

As a practical matter, an AMI system needs wireless communication mode to provide continuous, fast, and stable data transfer for optimizing data center operations within local areas as well as wide areas. Two wireless methods extensively used in an AMI system are Wi-Fi and an embedded system. These two methods are summarized below.

1) *Wi-Fi*: This is the most common technology in wireless communication. The implementation procedure in AMI is to establish wireless base stations that have to guarantee that the network provided is capable of covering all the IP-based meters. The base stations are connected with each other through the network bridge between different wireless access points. In addition, the choice of meter installation site is important. Before installing the wireless devices, a site survey of each installation location has to detect the strength of signal and interference around each meter to ensure the communication quality.

2) *Embedded System*: Compared with other wireless communication modes, an embedded system is a comprehensive method. Besides more stable information transmission, it allows users to combine multiple communication modes, based on their definite requirements, to complete the data transfer [164]. The embedded system can provide the optimal solution for wireless communication in an AMI system.

In addition, the global system for mobile (GSM), general packet radio service (GPRS), and 3G/4G networks could also be applied in an AMI system [165]. Since the exploitation of AMI systems, the option of advanced communication technologies has become increasingly complex. The implementers should consider more circumstances, such as reliability, operating maintenance, or capital spending to select the suitable communication mode. The communication of AMI relies on three networks: HAN, neighborhood area network (NAN), and WAN. The basic communication infrastructure of an AMI system is illustrated in Fig. 7.3.

Regarding a complete AMI system, the measured information that could be transmitted is not just electric power data; AMI could also gather the consumption information of water, gas, or heating. The main function of HAN is to integrate additional data and send it to relevant "smart" meters. Then, the data is sent through NAN, which is sometimes also identified as a LAN or field area network (FAN), to connect meters and concentrators. In the end, the WAN will connect the concentrator or the single "smart" meter to the head end system (HES), which has the ability to communicate with meters directly and also could be known as a meter control system. Between WAN and HES,

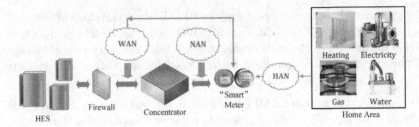

FIGURE 7.3
Communication infrastructure of an AMI system.

there will always be a firewall or special approval protocol to guarantee the security of data transmission.

7.1.3.3 Meter Data Management System (MDMS)

As the key component in the AMI system, the meter data management system (MDMS) performs long-term data storage and monitoring for the vast quantities of usage information delivered by IP-based meters. The data is typically imported to the MDMS first for preprocessing, e.g., verification, filtering, and disposing, before making it available for billing and analysis. Also, the MDMS can interact with some functional systems, such as a power-on-or-off control system or a dynamic prices system [166]. The working procedure of the MDMS is shown in Fig. 7.4. The MDMS can obtain a timely data report to do energy consumption forecasting, load capacities reporting, and customer service feedback [167].

Making full use of the gathered information is an important benefit of the AMI system. Apart from compiling the timely data, the MDMS can also maintain the data integrity even without data transfer, which means the MDMS can store the measured data under a network disconnecting situation. Currently, many electrical companies are planning to establish an MDMS based on their existing metering system, to promote work efficiency and performance.

7.2 Customer Information System

A customer information system (CIS) is a relational database that consists of a variety of customer-related information. The CIS always combines with the billing system. The electric utilities and companies should set an efficient billing rule for their customers once they deliver electricity to customers successfully. In addition, the CIS can be used to distinguish the profitability of customers through their energy consumption data. The CIS products

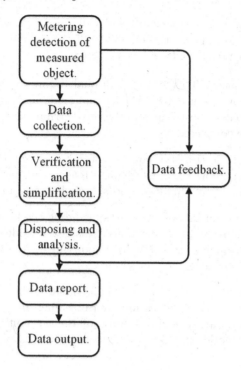

FIGURE 7.4
Flowchart of MDMS working process.

concentrate on basic connection, disconnection, meter servicing, and billing functions.

CIS multiple utility applications include:

- Meter Data Management (MDM)

- Work Management

- Consumer Experience Integrated Voice Response (IVR)

- Outage Management

- Marketing Programs

- Customer Relationship Management (CRM)

- Geographic Information Systems (GIS)

The detailed customer information, which includes the customer's profile, energy usage habits, electricity usage peak time, sales history, collections and receivables, and metering data, should be recorded by the CIS. The distribution management system sometimes can only obtain the SCADA data; the customer information may not be available locally in the SCADA/DMS control center. In order to improve the reliability of the system, a summary of high-level generic features for CIS is listed as [168]:

- Customer information: This is the relational database that consists of customer data, such as account number and type, and association with substation and feeder; most importantly, the distribution transformer IDs all consumers are connected to.

- Historical metering statistics: Frequency of metering, manufacturer, type and model, ratings, power factors, manual reading history, such as value and date, as well as maintenance records, including calibration.

- Customer relationship: List of customers including the history of connections, disconnections, and reconnections. This also includes the records of first time connection, disconnection due to non-payment, and reconnection with dates that were scheduled for service.

- Invoices to consumers: Corporate/ individual name, address, account type, current usage, tariff schedules, reading date, billing date, due date, and aged debt. This database includes most of the accounting information that can include notice of disconnection if applicable and the dates of disconnection.

- Payment system: This payment system includes cash register, checks and bounced checks, payment structure, such as prepaid amounts and partial payments, payment locations, receipt archive, and credit history.

When the operator or crew search for the potential fault location, the customers' feedback is important (e.g., trouble call tickets). Therefore, the customer information should be integrated into the complete data package. There is a synchronization between the internal outage management and the external customer information system. The outage management system interacts with the physical device throughout the entire feeder, which usually relies on the operational technology (OT), while the CIS mainly depends on the information technology (IT).

7.3   Trouble Call Tickets

Trouble call tickets are an acknowledgment by the customer billing center about recent power outages complained of by affected individuals. These tickets are traditionally collected over the phone where an association of the customer ID with the distribution transformer ID is made. Presumably, the incoming outage tickets can be related to one another. During stormy weather conditions, the affected outage can be feeder-wide. Establishing a timely report of trouble call tickets is essential because it can (1) help operators to identify outage, (2) confirm a large scale outage, (3) enhance the search for a major outage, and (4) identify independent events.

The collaborative integration of trouble call management with DMS has recently attracted considerably wide attention in power enterprises. Along with the direct benefits of coordination for a out-of-service event in dispatch, trouble call management can also relate to marketing and production in other departments of a utility company. As a search for a fault can be time-consuming and challenging, the trouble call management can relate to customer repairing calls, real-time network monitoring, and repair work management. These systems may still be in the state of dispersing deployment and work independently as applications to outage management. Repetitive repair calls may not directly associated with an outage event.

A typical scenario of a service restoration cycle is as follows: When an outage occurs, the dispatchers are in the process of trying to answer the telephone as well as call out for additional assistance. Trouble calls are manually entered on forms, with the name of the customer, address, telephone number, outage problems, etc. The customer billing center can identify the location of customers and associate these complaints with the power outage and potential common problems with other customers in a geographical region. Using a map with dynamic information on the energization state, the operator may be able to find out potential fault segments based on their engineering training and experience. Once the estimated area is identify, the information will be passed on to crew members at the site to investigate the exact fault location as well as finding the root cause of the problem. Reports from the crew members are coordinated by the operators to assign them a job to repair damage as well as to restore power [169].

7.3.1   Trouble Call Management System

Trouble call management (TCM) is an application information platform to handle the process of receiving trouble calls and making repairs during unscheduled management. The functional modules of TCM are integrated into DMS and CIS [170]. The overall structure of TCM consists of three layers and multiple modules:

- **Application Layer:** The applications include fault call management, fault auxiliary analysis, deploying of repairing resources and work management, evaluation, and optimization analysis.

- **Data Layer:** This data includes the call information, fault information, repairing information, work orders, deploying of repairing resources, and an expert database.

- **External Data Layer:** This layer is not a part of the TCM, but there are always data interactions between the two of them. The modules in this layer include the power consumption data acquisition system, DMS CIS, SCADA, summarized production information, GPS, and emergency commands.

On average, about ten low-voltage single customer service interruption cases and two low-voltage multiple customer service interruption cases are reported per day. Table 7.2 shows a summary of various service interruption problems in a service area of a district [171]. There was about one higher voltage customer service interruption case every three days and one feeder trip incident per week. The single customer service interruption cases are usually due to Watt meter, switch contact, service line, or fuse problems. For low-voltage multiple customer service interruption cases, the primary causes are fall of the low-voltage lines, burnout of fuses installed in the primary side of the distribution transformers, transformer overloads, and deterioration of transformer bushings, etc. It normally takes 30 to 40 minutes to restore the service of low-voltage customers, 50 to 60 minutes to restore high-voltage customer service, and four hours for a feeder trip case.

One commonly used method for trouble call analysis is the implementation of a visual representation of the geographical map. Each region of a map is divided into multiple polygons as "polygoning [172]." Each polygon has a distribution transformer (DT). By analyzing a customer complaint call, the association with the distribution transformer and customer ID will be matched. This can be an independent event, but the timing of incoming calls are important to relate events to one another. For a more widely implemented method, every incoming trouble call is associated topologically with their relationship to DT ID and the sequence of tickets when they are issued. Under the same timeframe, the upstream tracing method is utilized. Upstream tracing from the location of the callers is then conducted to determine the common devices for a relevant outage. This requires the GIS topology to be up to date, and specific protective relaying devices are well defined in the network. Such a method can be more conclusive than the polygoning method due to the nature of the spatiotemporal correlation. The drawback of this method, though, is the uncertainties involved in locating the probable cause that is completely independent of others (assuming there are multiple events occurring within a longer timeframe, e.g., within a 4-hour interval) [173–176].

7.3.2 Fault Localization Incorporating Trouble Calls

In this section, 5 scenarios will be illustrated to explain trouble call analysis combined with FI-based fault localization.

Scenario 1

Scenario 1 is an ideal case. As shown in Fig. 7.5, we zoom in on a lumped

TABLE 7.2: Summary of Service Interruption Events.

Service Interruption		Causes	Average Time to Restore Service
Low-voltage area (110 V, 220 V, 380 V)	Single customer 10 cases per day	Problems at watt meters, switch contacts, service line, fuses, etc.	30–40 min
	Multiple customers 2 cases/day	Fuse burnout, transformer overload, fall of low-voltage line, etc.	40–50 min
High-voltage area (11 kV and 22 kV)	High-voltage customer 1 case/3 days	Deterioration in transformer, switch, fuse, etc.	50–60 min
	Feeder trip 1 case/week	Feeder overload, unbalance, short circuit, deterioration in CT, PT, capacitor, etc.	4 hours

load with the distribution transformer and households in a feeder. When a fault occurs, the breaker at the feeder head will be tripped and the corresponding fault indicators will display the detection results. In this case, the trouble call tickets are used to confirm the outage events. Both the trouble call tickets and the breaker tripped would be timestamped. Combined with the fault segment localization described previously, the trouble call tickets can narrow down the search area and provide the potential fault that is possible in the low-voltage area (fuse, transformer, meters, lines, etc.). In this scenario, the trouble call tickets narrow down the potential fault area to e_6 (medium-voltage area) and e_8 (low-voltage area).

Under the ideal scenario, the association of incoming events falls within a time interval that is consistent with an event occurrence of the disturbance, the breaker/recloser is sensitive enough to trip, fault indicators are rendered from the FRTU on time, and trouble call tickets are issued minutes/hours later to reaffirm the outage association from the SCADA database.

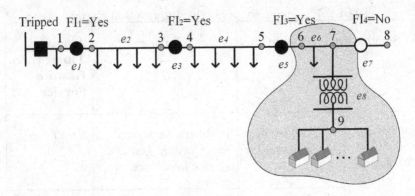

FIGURE 7.5
Ideal case of trouble call tickets analysis.

Scenario 2

Under realistic scenarios, the breaker/recloser might not be sensitive enough to trip, fault indicators may not be correctly rendered from the FRTU, and incoming measurements (referred to as FIs) are delayed, lost, or missing or do not fall within the time interval. If the recloser/breaker does not trip, the fuses may burn with an electrical fault closer to the distribution transformer(s), resulting in a power outage to those customers. Trouble call tickets take longer to collect. Presumably, the majority of customers do not have smart meters set up to report their consumption in a higher data resolution.

As shown in Fig. 7.6, only one household is connected to a distribution transformer. If there are no smart meters under this circumstance, the customer information can only harness to the SCADA or CIS. If this customer is experiencing a power outage caused by a fault (fuse burned, deterioration in the transformer, or a fault on the customer side), the DMS and operators cannot detect this outage since the fault current is too small to cross the FI's threshold. All FIs in the terminal units display "No" in this case. The only way to localize the fault is to rely on the trouble call tickets. In this scenario, the potential fault area can be inferred as e_6.

Scenario 3

As shown in Fig. 7.7, in this scenario, there is no primary distribution communication; therefore, the breaker is not tripped, but two FIs show "Yes" within the feeder. Zoom in on one of the lumped loads within the faulted section. A distribution transformer connects with 3 households. Under common circumstances, the operator or crew will consider checking the line segment between "Yes" and "No" FIs first and then exclude the redundant part (the

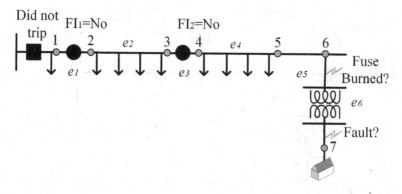

FIGURE 7.6
Realistic case of trouble call tickets analysis with a one-transformer-one-customer case.

segment between two "Yes" FIs). However, an event of the trouble call tickets issued coming from the line segment between two "Yes" FIs is probably caused by fuse burnout. This situation might be one of the considerations for a fault in the surrounding area (indicated by a cluster). Either of these two phone calls to the CIS will be treated as a trouble call issue. Compared with the fault segment localization, additional trouble call tickets actually provide additional information to narrow down the search area and let operators consider the whole potential faulted area. In this case, it is hard to tell whether there is a single event or two independent events. Without the trouble call tickets, the potential faulted area includes e_2, e_3, e_5, and e_7, but the crew only needs to check e_5 and e_7 with the trouble call tickets analysis.

Scenario 4

Scenario 4, which is shown in Fig. 7.8, is similar to Scenario 3: The breaker is not tripped and one FI response is abnormal. In addition, the fuse of the zoomed area is burned. Therefore, there are two independent events. If FI_2 also shows "Yes," the detection result will be reasonable. This situation is possible since the FI catches the transient current information and can generate an incorrect result temporarily. Relatively, the trouble call tickets are more reliable because they come from customers' genuine outage responses. The crew will detect the trouble call area first and then check the changes of FIs.

Scenario 5

As shown in Fig. 7.9, the breaker did not trip and all FIs show "No." The

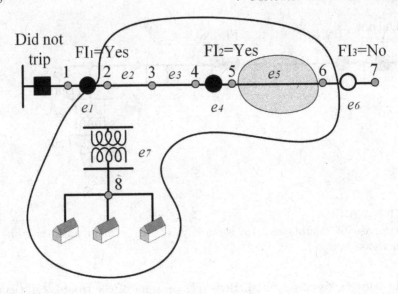

FIGURE 7.7
Realistic case of trouble call tickets analysis with three customers.

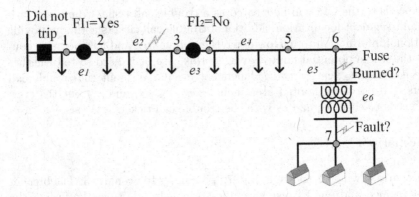

FIGURE 7.8
Realistic scenario of trouble call tickets analysis with three customers.

trouble call tickets sequence and the timestamps of each call, in this case, are important, since they can help operators judge if there is a primary network fault or not. If the time interval between two trouble calls is long enough (more than Δt, which is the presetting time interval for a fault identification), the fault can be treated as the secondary network fault only. Otherwise, the primary network fault localization can be inferred according to the sequence of trouble call tickets. This content will be discussed in the next section in detail. In this scenario, the time interval between these two trouble calls from customer 3 and customer 6 is more than 4 hours. Therefore, there are two independent events occurring in the secondary network.

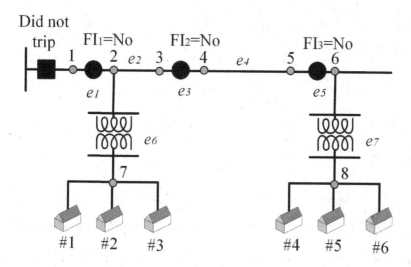

FIGURE 7.9
Realistic scenario of trouble call tickets analysis with six customers.

Detailed information of Scenarios 1, 2, 3, and 5 with corresponding input variables are shown in Listing 7.1. The MATLAB sample code illustrated in Listing 7.2 demonstrates the fault localization from the trouble call tickets. The input variables include the topological incidence matrix, the connection states, the source vector, and the index of FIs.

Listing 7.1
Input Variables of 4 Scenarios for Fault Localization Analysis Based on Trouble Call Tickets

```
1  casenumb = input('Input the case number (1-4): ');
2
3  if casenumb == 1
4  IncM = [1 0 0 0 0 0 0 0 0
5          1 1 0 0 0 0 0 0 0
6          0 1 1 0 0 0 0 0 0
7          0 0 1 1 0 0 0 0 0
8          0 0 0 1 1 0 0 0 0
9          0 0 0 0 1 1 0 0 0
10         0 0 0 0 0 1 1 1 1
11         0 0 0 0 0 0 1 0
12         0 0 0 0 0 0 0 1];
13
14 E_D = [1 1 1 1 0 1 1 1];
15 V_D = [1 0 0 0 0 0 0 0 0];
```

```
16  FI = [1 3 5 7];
17  Trouble = input('input the node number of trouble call ticket: ...
       '); %9
18  Burned = find(IncM(Trouble,:)==1);
19  E_D(Burned) = 0;
20
21  elseif casenumb == 2
22  IncM = [1 0 0 0 0 0
23          1 1 0 0 0 0
24          0 1 1 0 0 0
25          0 0 1 1 0 0
26          0 0 0 1 1 0
27          0 0 0 0 1 1
28          0 0 0 0 0 1];
29
30  E_D = [1 1 1 1 1 1];
31  V_D = [1 0 0 0 0 0 0];
32  FI = [1 3];
33  Trouble = input('input the node number of trouble call ticket: ...
       '); %7
34  Burned = find(IncM(Trouble,:)==1);
35  E_D(Burned) = 0;
36
37  elseif casenumb == 3
38  IncM = [1 0 0 0 0 0 0
39          1 1 0 0 0 0 1
40          0 1 1 0 0 0 0
41          0 0 1 1 0 0 0
42          0 0 0 1 1 0 0
43          0 0 0 0 1 1 0
44          0 0 0 0 0 1 0
45          0 0 0 0 0 0 1];
46
47  E_D = [1 1 1 0 1 1 1];
48  V_D = [1 0 0 0 0 0 0 0];
49  FI = [1 4 6];
50  Trouble = input('input the node number of trouble call ticket: ...
       '); %8
51  Burned = find(IncM(Trouble,:)==1);
52  E_D(Burned) = 0;
53
54  elseif casenumb == 4
55  IncM = [1 0 0 0 0 0 0
56          1 1 0 0 0 1 0
57          0 1 1 0 0 0 0
58          0 0 1 1 0 0 0
59          0 0 0 1 1 0 0
60          0 0 0 0 1 0 1
61          0 0 0 0 0 1 0
62          0 0 0 0 0 0 1];
63
64  E_D = [1 1 1 1 1 1 1];
65  V_D = [1 0 0 0 0 0 0 0];
66  FI = [1 3 5];
67  Trouble = input('input the node number of trouble call ticket: ...
       '); %[7 8]
68  for i = 1:length(Trouble)
```

```
69        j = Trouble(i);
70        Burned = find(IncM(j,:)==1);
71        E_D(Burned) = 0;
72   end
73
74   end
75   fault = find(trobfloc(IncM,E_D,V_D,FI)==0);
76   fprintf('the fault occurs in %d\n',fault);
```

Listing 7.2
Algorithm of Fault Localization Analysis Incorporating Trouble Call Tickets

```
1   function line_state = trobfloc(incMat, EdgeVec, NodeVec)
2
3   % Convert incidence matrix to adjacency matrix.
4   YesNoMat = diag(EdgeVec);
5   incMatr = incMat * YesNoMat;
6   adjMat = incMatr*incMatr';
7   adjMat = abs(adjMat - diag(diag(adjMat)));
8
9   % Initialize values.
10  affectnode_vec = NodeVec;
11  affectnode_old = affectnode_vec*0;
12
13  % Criteria that if the result is equal to the previous ...
          iteration result.
14  while (length(find(affectnode_old == 0)) - length(find ...
          (affectnode_vec == 0))) ≠ 0
15      affectnode_old = affectnode_vec;
16      % Multiplication based on the topology (Adjacency Matrix).
17      affectnode_vec = affectnode_old * adjMat + affectnode_old;
18      idx = find(affectnode_vec > 0);
19      affectnode_vec(idx) = 1; % Replace all non-zero elements ...
              to 1s.
20  end
21
22  % Find affected connections based on affected nodes from ...
          incidence matrix.
23  line_state = ones(1,length(incMat(1,:)));
24  node_ind = find(affectnode_vec == 0);
25  for i = 1:length(node_ind)
26      j = node_ind(i);
27      line_ind = find(incMat(j,:) == 1);
28      line_state(line_ind) = 0;
29  end
30  %line_state(FI) = 1;
```

7.3.3 Potential Fault Segment Identification Based on Multiple Trouble Calls

Because a node is connected to a distribution transformer with a line segment in the primary network and customers also connect to the distribution transformer, a potential fault that occurs in the primary network can also be inferred by trouble call tickets. If the time interval between two trouble calls is limited and they are coming from two different branches or lumped loads, the probability that the potential fault exists in the primary network is higher than that it may be in the secondary network. The trouble calls' incoming sequence will help operators narrow down the faulted section.

7.3.3.1   One Simple Feeder

Fig. 7.10 (b) demonstrates the graph representation of a single feeder topology. A detailed diagram is shown in Fig. 7.10 (a), where each node is connected with at least one distribution transformer and the transformer is connected with multiple households. In this situation, the breaker is not tripped and all of the FIs in this feeder display normal results. Assume the presetting time interval for a fault identification Δt is 4 hours.

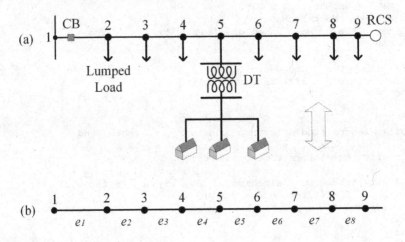

FIGURE 7.10
A comparison between a single feeder topology and its graph representation.

As shown in Fig. 7.11, the CIS recorded the first trouble call coming from node 5 at t_1 (1st hour). If there is a primary network fault at this time, it most likely exists in e_4 since no other reference trouble calls came in. After 2 hours, the CIS received another trouble call from node 8 at t_2 (3rd hour). According to this information, the fault still possibly in e_4 since node 4 is nearer to the injection source. However, if the trouble call from node 8 comes first, the inference of the first step is to check e_7 rather than e_4 because e_7 connects with node 8. In the ideal case, if a fault occurs in e_4, all customers from nodes 5 to 9 should call in to complain about their situations, but not everyone is highly motivated to make a phone call immediately after the power outage in a real scenario. At t_3 (4th hour), another trouble call comes from node 2; then the potential faulted segment is inferred as being in e_1.

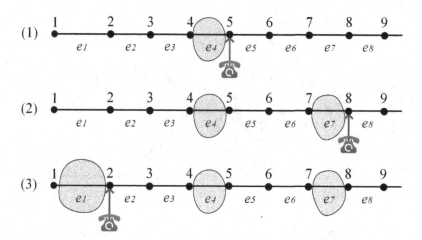

FIGURE 7.11
Potential fault segment identification based on multiple trouble calls in a single feeder.

If the Δt is adjusted as 3 hours, the fault identification process will only consider the first two trouble calls, since the last one exceeds the time interval. Therefore, the timestamp and sequence of each trouble call are important for potential fault inference.

7.3.3.2 One Feeder Branches Out with Two Sub-Feeders

The inference process in one feeder branches out with two sub-feeders similar to the one simple feeder case. As shown in Fig. 7.12, assuming a trouble call is from node 4 at timestamp t_1, the crew will check the segment e_3 if no other information comes in. Within the Δt interval, another trouble call comes from node 3 and is received by the CIS; therefore, the potential fault section is changed to e_2.

Likewise, as shown in Fig. 7.13, if the first trouble call is from node 7, the first inference result is e_6, while another trouble call is from node 4 within Δt. The potential fault segment should be e_3, since node 4 is closer to the power source.

The MATLAB sample code illustrates the algorithm of the potential fault segment identification based on multiple trouble calls, and the two scenarios discussed above are shown in this script. In summary, multiple trouble call tickets are capable of exploring the topologies from the input of 1) relying heavily on trouble call tickets issued over time, and 2) using fault indicator(s) and breakers not tripped as the references.

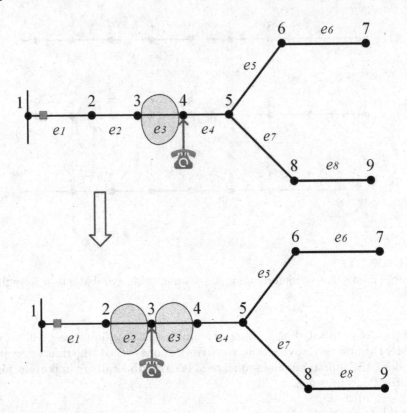

FIGURE 7.12
Potential fault segment identification based on multiple trouble calls in a single feeder with two branches.

Listing 7.3
Potential Fault Segment Identification Based on Multiple Trouble Calls

```
1  clear
2  clc
3
4  Δ_t = 4; % number of hours.
5  start_t = 720; % 12:00. 1-1440 minutes. e.g. 360 means 6:00; ...
       720 means 12:00.
6  start_hour = floor(start_t/60);
7  start_min = start_t-start_hour*60;
8
9  casenumb = input('Input the case number (1-2): ');
10
11 if casenumb == 1
12     IncM = [1 0 0 0 0 0 0 0
13             1 1 0 0 0 0 0 0
14             0 1 1 0 0 0 0 0
15             0 0 1 1 0 0 0 0
16             0 0 0 1 1 0 0 0
```

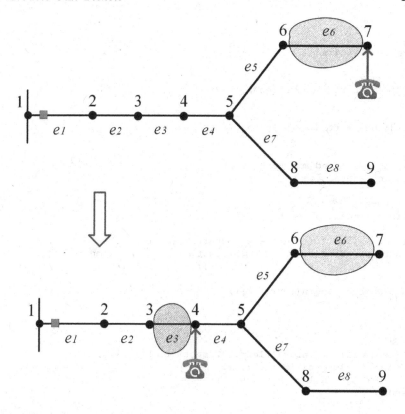

FIGURE 7.13
Potential fault segment identification based on multiple trouble calls in a single feeder with two branches.

```
17                 0 0 0 0 1 1 0 0
18                 0 0 0 0 0 1 1 0
19                 0 0 0 0 0 0 1 1
20                 0 0 0 0 0 0 0 1];
21
22       V_D = [1 0 0 0 0 0 0 0 0];
23
24   elseif casenumb == 2
25       IncM = [1 0 0 0 0 0 0 0 0
26               1 1 0 0 0 0 0 0 0
27               0 1 1 0 0 0 0 0 0
28               0 0 1 1 0 0 0 0 0
29               0 0 0 1 1 0 1 0
30               0 0 0 0 1 1 0 0
31               0 0 0 0 0 1 0 0
32               0 0 0 0 0 0 1 1
33               0 0 0 0 0 0 0 1];
34
35       V_D = [1 0 0 0 0 0 0 0 0];
36   end
```

```
37
38   % Find the branch node
39   Bnode = find(sum(IncM,2)>2);
40
41   % Convert incidence matrix to adjacency matrix.
42   AdjM = matrixconvert(IncM,1);
43
44   % Find the nodes connected with the branch node.
45   Bsubnode = find(AdjM(Bnode,:)==1);
46
47   % Delete the connected node before the branch node.
48   headnode = find(min(Bsubnode));
49   Bsubnode(headnode) = [];
50
51   % Input timestamp for trouble call tickets.
52   Timestamp_all = input('Input trouble call tickets timestamps: ');
53
54   start = find(Timestamp_all ≥ start_hour, 1 );
55   endtime = find(Timestamp_all≥start_hour & Timestamp_all ≤ ...
           start_hour + ∆_t, 1, 'last' );
56
57   NodeNumb = length(IncM(:,1)); % Number of nodes of the topology.
58
59   % Initialization of the trouble call vector with largest node ...
           index.
60   % (Preassume all trouble calls are from the longest node)
61   Trouble_call = zeros(1,length(Timestamp_all));
62   for m = 1:length(Timestamp_all)
63       Trouble_call(m) = NodeNumb;
64   end
65
66   % Input corresponding node number of trouble call ticket based ...
           on timestamps.
67   connect_change = ones(1,length(IncM(1,:)));
68   E_D = ones(1,length(IncM(1,:))); % Initialization of ...
           connection state.
69   for i = 1:length(Timestamp_all)
70       j = Timestamp_all(i);
71       Trouble_call(i) = input('Input the node number of trouble ...
               call ticket:');
72       E_D = ones(1,length(IncM(1,:))); % Reconnecte connection ...
               state.
73
74       if j≥start_hour && j ≤ start_hour + ∆_t
75           Trouble = min(Trouble_call(start:endtime)); % Find the ...
                   node index closer to the power source.
76           Fault = find(IncM(Trouble,:)==1,1,'first'); % Find the ...
                   injection edge connected with the node.
77           E_D(Fault) = 0; % Mark the edge.
78
79           if ismember(Trouble_call(i),Bsubnode)≠0
80               % The special situation: two timestamps and trouble ...
                       calls from two sub-branch nodes
81               if j≥start_hour && j ≤ start_hour + ∆_t
82                   Trouble = Trouble_call(i);
83                   Fault = find(IncM(Trouble,:)==1,1,'first'); % ...
                           Find the injection edge connected with the node.
```

```
84                  E_D(Fault) = 0; % Mark the edge.
85              else
86                  continue
87              end
88          end
89      else
90          continue
91      end
92      connect_change = [connect_change;E_D]; % Record the ...
            connection state change.
93  end
94  connect_change
95
96  iter = length(connect_change(:,1));
97  fault = find(connect_change(iter,:)==0);
98  fprintf('The fault occurs in segment %d\n',fault);
99
100 if start_min < 10
101     fprintf('The start time is %d:0%d and Δ_t is %d ...
            hours\n',start_hour,start_min,Δ_t);
102 else
103     fprintf('The start time is %d:%d and Δ_t is %d ...
            hours\n',start_hour,start_min,Δ_t);
104 end
```

7.4 Outage Escalation

An outage escalation is to infer the potential fault location from multiple trouble calls over a period of time. Operators coordinate at a control center with 1) the crew who might already be at the site, 2) the location of the fault segment within the boundary based on fault localization and isolation, 3) how many customers are affected, 4) the outage state; that is, whether the outage is scheduled or unexpected (unscheduled), and 5) CIS with incoming trouble calls. This issue can relate to the optimization of the traffic conditions in searching for potential faults and the availability of crew members at sites. Sometimes, the search space can be tens of miles square and it will take an uncertain amount of long hours to pinpoint the faulted segment of a line. This remains an emerging issue and the optimization problems can be subject to, the overhead lines where may intercross with some roads that do not align parallel to the road maps.

7.5   Conclusions

The trouble call tickets confirm an outage where it may be associated with other disturbance events. Hence, the relation of customers-to-distribution transformers is critical to connect with a computerized management system. Trouble call tickets can be replaced with smart meters if the utility implements IP-based meters at all their customer sites. This chapter visits the technologies of AMR/AMI and how they could help to relate the fault events from the sequential order of reporting to a power outage at the secondary level of a distribution network. Depending on the priority of customers, the escalation of the search can start at the root node nearest to the complaining customers. The trouble call tickets can also be related to other independent events that can occur during times of abnormal severe weather.

There are emerging subjects that relate to smart meters where the reporting of outages to the distribution control center (DCC) can be rapid. Such situations will help dispatchers in the DCC identify the potential outage area precisely, and infer the root cause of the potential outages. The disadvantage of relying on a trouble call reporting system is that if a fault occurs at a distribution transformer (DT) with only one residential house connecting to it, the resident must make an effort to report the problem almost immediately. If not, the outage of the customer will not be escalated until the customer reports it. Also, most reports of associated outages by customers can be subject to their willingness to take the initiative and report their outage to the utilities.

The customer billing center often relates the distribution transformer (DT) identifier to the customer billing information. The sharing of such information with the DCC helps to identify the connected customers to the various locations of the laterals, which can be directly referred to the immediate root node of a feeder as a potential root cause of the problem. This will simplify the search space, help to reduce outage time, and improve overall reliability.

Mini Project 6: Integrate with Fault Isolation and Partial Restoration Module

The MATLAB script, provided in this chapter are the inference of potential locations of fault. This fault area may happen in the secondary network. It can also be in the primary network of a distribution network. The integration of this chapter and the last chapter is critical. Assuming it is a single event disturbance, the integration of this and the last chapters will provide the best potential faulted area(s), which can be useful and help shorten the search time. Although sometimes the search area can be large, the trouble call tickets will confirm the outage and may help dispatchers to best guests the fault location, which can be useful information.

8

Outage Management

The previous chapters have visited the topics relevant to searching for electrical faults in which the exact location of the short circuit is not defined. The large size of a faulted area may affect hundreds of customers. There remains a possibility that we still may be able to isolate the faulted segment only with the involvement non-remote-controlled switches. By doing so, the majority of customers may be able to have power during the repair period. This chapter establishes an update incidence matrix that only consists of non-remote-controlled switches. It explores the switching steps by narrowing down the smallest area. Each attempt can involve bisection search. The reliability indices, such as SAIDI, SAIFI, or CAIDI, can be calculated using the incidence matrices. Statistically speaking, 80% of power outages occur in an electrical distribution network [177]. Therefore, in power system design and operation of the distribution network, a reliability assessment is one of the most important factors. Thus, it is necessary to identify critical components for outage scheduling, and to hedge power systems against potential risks [178].

8.1 Outage Management System

The outage management system will have information about the reliability of the distribution network. Outage management strives to coordinate the following modules:

- **Crew scheduling**: It is used to identify available crews in a scheduled or unscheduled time duration.

- **Switching steps**: This module interacts with the unbalanced sweeping technique of power flow. It can be divided into:

  - automatic switching generation coming from the fault localization, isolation, and service restoration, and

  - manual switching generation heuristic from user's and crew's experiences.

- **Trouble call system**: This could be related to scheduled or unscheduled outages. Before a scheduled outage, a staff member will notify the customers in the outage area, to avoid redundant trouble call tickets. If the trouble call is from the unscheduled outage area, the customer trouble call center will record the account number and the location of the reported outage to coordinate with the fault management system, the switching, and the crew scheduling modules.

- **Customer information system**: This module has the complete information for all customers, which includes the addresses, consumption information, payment histories, etc., to coordinate with other modules.

The input data of the outage management system to determine the root cause of a power outage includes:

- Up-to-date GIS topological datasets to manage and understand the latest topological statuses of the distribution network.

- SCADA measurement points, which contain:

    - Fault indicators with "Yes" or "No" indications;

    - Switch statuses with the binary measurements from FRTU/FTU;

    - Loading conditions with analog measurements.

- Trouble call tickets or smart meters from the secondary network, to show the real-time information (No kWh data within 15 minutes or longer).

8.2   Crew Coordination

Multiple power outages in a large-scale distribution system could be caused by significant damage and faults. Under this condition, many service zones in a large-scale area with breakers or reclosers are tripped simultaneously. The electric companies and utilities need to send field crews to repair the damage and clear the outages as soon as possible. Due to the limited number of crew resources in large-scale multiple contingencies, the effectiveness of the crew assignment and management is especially crucial.

Traditionally, the electric failure in a distribution feeder can be detected by the circuit breaker (CB) or the fault indicator (FI) on the feeder head. Also, the outage or abnormal event may be responded to by trouble calls from customers or by the fault report from utility crews according to field inspection. The failure event can be found much more easily in a modern feeder with the automation system as discussed above. However, whether in a traditional system or a modern automatic system, power restoration in the faulted areas can be recovered only after the completion of the repair [179].

Electric utilities and companies will send their crew to the field to localize and confirm the extent of the damage and pass the information to the dispatching control center (DCC). The event report and log from field crews should include the fault location, type of event, the resources of the crew, and the tools, vehicles, and materials needed for the repair [179, 180]. The estimated time needed to repair the event and restore the power should be provided to the DCC.

A crew management system should be capable of dealing with the voluminous data and complicated situations in multiple outages. How to coordinate the crew database, the vehicle database, and the database of repair parts/-components is the core of the intelligent resource management system. The event information in this system is received and recorded from the field crews. In addition, the information about spatial mapping and feeder facilities of the distribution system, which includes the topological information of feeder configuration and the data of feeder facilities and customers, are provided by the database of the outage management system. A historical event database will be created for the failure events once the repairs are completed. Ultimately, the crew management system will assist the dispatchers and operators by tracking the condition of the events and the resources of crews, and by proposing an effective assignment of the field crews and tools for the events. The two major components of the crew management system are [180]:

- **Crew scheduling and assignment:** This component is capable of scheduling and assigning crews from all sites. According to the changeable timetable and real situations, an optimized assignment of crews can be provided. The reference data can be obtained from the mainframe computer and combine crew scheduling experience, labor rules, equipment arrangement, etc. AI-technique is the core module in this component.

- **Crew management and tracking:** This component is able to implement the coordinated crew schedule for each site and accommodates changes due to the absence or late arrival of crew members. The reference data in this component is collected from the traditional databases and is fully integrated with the crew scheduling and assignment component.

A detailed example demonstrates the reliability of a distribution system for fault segment localization and crew coordination within the located fault segment. As shown in Fig. 8.1, in the first stage, the entire feeder is experiencing the power outage as highlighted in red (see color e-book). The circuit breaker at the feeder head is tripped (indicated as an empty box) since a fault occurs somewhere within the feeder. The two fault indicators installed in the remote-controlled switches (black spots) show the results as "Yes" and "No," respectively. The tie switch represented by an empty circle is also shown as "No." The fault must exist between the "Yes" and "No." The distribution control center can perform the fault segment isolation and temporary restoration within 10 minutes so that the stage will convert to stage 2.

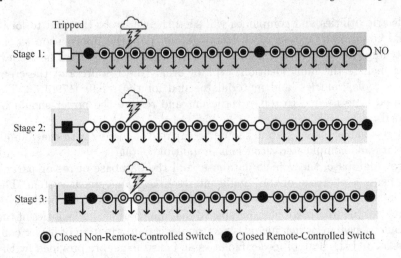

FIGURE 8.1
Fault segment localization and crew coordination within the located fault segment in a distribution system (4 remote-controlled switches including CB; the rest is not remote-controlled).

In stage 2, the tripped breaker is reclosed, the two remote-controlled switches at both ends of the faulted segment are opened, and the tie switch is closed. The energized area is highlighted in light gray. However, more than half of the customers in this feeder are still unhappy, since the de-energized area is significant and large. Therefore, the operator and crew want to narrow down the zone to the smallest area so that a minimum of customers will be affected. In this case, as shown in stage 3, only a single lumped load is de-energized.

In terms of reliability, stage 2 will be converted to stage 3, and it will take around 2 to 5 hours before the topology gets back to the normal stage (stage 1). However, the operator has to work closely with the field crew who are at the site to figure out the strategic plan to begin the narrow process. Usually, the crew will apply the bi-section search to perform the relative operations. In summary, the conversion from stages 1 to 2 is controlled by the remote-controlled switches automatically, while the stages from 2 to 3 relate to the coordination between operator and crew. The whole process from when the fault occurred to the normal stage will take tens of hours.

A detailed crew coordination process between stages 2 and 3 is illustrated in Fig. 8.2. After the control center checks the crew schedule and finds the closest one, the bi-section search can be performed in the faulted line segment. The search starts from the midpoint of the line segment. The crew opens the non-remote-controlled switch and asks the control center to try to close the remote-controlled switch at the feeder head. When they close the switch, if the fault is still within the front region, the breaker will be tripped immedi-

ately. Topologically, the fault can be located in the subsystem highlighted in
blue (see color e-book). The crew then can utilize the zone by the zone-rolling
method to open the nearest adjacent switch, or continue the bi-section approach, which is applied in this case. The crew repeats this process to open
the middle switch in the remaining area, to eventually localize the smallest
region highlighted in green (see color e-book). This search process contains
3 iterations. If applying the zone-by-zone rolling method, this process will
extend to 5 iterations in this case. However, the fault occurs randomly; the
method chosen can only rely on the operator and crew's experience.

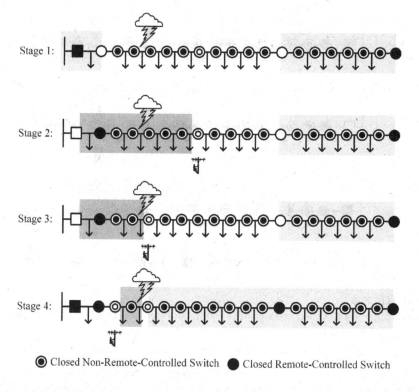

FIGURE 8.2

An example of a detailed crew coordination process based on bi-section search
method.

Fig. 8.3 shows another example. After the fault segment localization and
isolation, the fault exists in the tail part of the feeder. In this case, the zone-by-
zone rolling method is applied to narrow down the fault area. The crew tring
to open the non-remote-controlled switch start from the first one and then
go on to the next. During each operation, the operator will close the remote-
controlled switch and check if the breaker is tripped. After 3 iterations, the
smallest faulted region can be identified, and the closest non-remote-controlled
switch has to be opened again by the crew team.

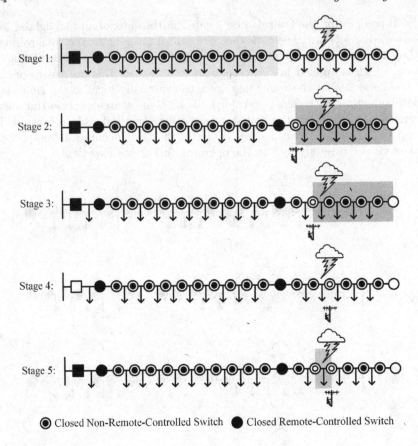

Stage 1:

Stage 2:

Stage 3:

Stage 4:

Stage 5:

◉ Closed Non-Remote-Controlled Switch   ● Closed Remote-Controlled Switch

FIGURE 8.3
An example of a detailed crew coordination process based on the zone-by-zone
rolling search method.

Figs. 8.4, 8.5, and 8.6 illustrate the typologies of the three cases for MAT-
LAB simulation. Detailed information on these three cases with corresponding
input variables is shown in Listing 8.1. The MATLAB sample code illustrated
in Listing 8.2 demonstrates a crew coordination management based on the location of the field crews. The input variables include the topological incidence
matrix, the connection states, the source vector, and the index of switches.
The fault localization function is also utilized. The basic idea of this algorithm
is to dispatch the closest crew to locate and repair the faulted section.

Listing 8.1
Examples of Input Variables for the Crew Coordination Algorithm

```
1  casenumb = input('Input the case number (1-3): ');
```

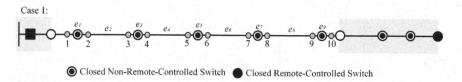

FIGURE 8.4
Topology for MATLAB simulation case 1.

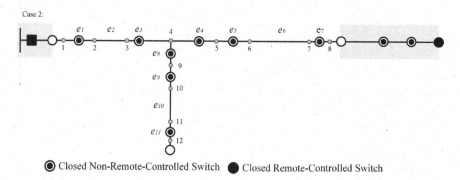

FIGURE 8.5
Topology for MATLAB simulation case 2.

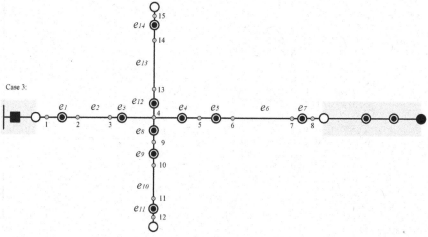

FIGURE 8.6
Topology for MATLAB simulation case 3.

```
2
3   %-----------------------------------------------------------%
4   % Case 1: 10-nodes & 9-edges example.
5   %-----------------------------------------------------------%
6   if casenumb == 1
7
8   IncM =[1 0 0 0 0 0 0 0 0
9          1 1 0 0 0 0 0 0 0
10         0 1 1 0 0 0 0 0 0
11         0 0 1 1 0 0 0 0 0
12         0 0 0 1 1 0 0 0 0
13         0 0 0 0 1 1 0 0 0
14         0 0 0 0 0 1 1 0 0
15         0 0 0 0 0 0 1 1 0
16         0 0 0 0 0 0 0 1 1
17         0 0 0 0 0 0 0 0 1];
18
19  connect_sta =[1 1 1 1 1 1 1 1 1];
20  sourceVec = [1 0 0 0 0 0 0 0 0 0];
21  nrcsw = [1 3 5 7 9];
22
23  % The fault occurs randomly.
24  N = length(IncM(1,:));
25  fault_loc = randsrc(1,1,2:2:N);
26  % The fault location options are [2 4 6 8];
27  end
28
29  %-----------------------------------------------------------%
30  % Case 2: 12-nodes & 11-edges with a branch example.
31  %-----------------------------------------------------------%
32  if casenumb == 2
33
34  IncM = [1 0 0 0 0 0 0 0 0 0 0
35          1 1 0 0 0 0 0 0 0 0 0
36          0 1 1 0 0 0 0 0 0 0 0
37          0 0 1 1 0 0 1 0 0 0
38          0 0 0 1 1 0 0 0 0 0
39          0 0 0 0 1 1 0 0 0 0 0
40          0 0 0 0 0 1 1 0 0 0 0
41          0 0 0 0 0 0 1 0 0 0 0
42          0 0 0 0 0 0 0 1 1 0 0
43          0 0 0 0 0 0 0 0 1 1 0
44          0 0 0 0 0 0 0 0 0 1 1
45          0 0 0 0 0 0 0 0 0 0 1];
46
47  connect_sta =[1 1 1 1 1 1 1 1 1 1 1];
48  sourceVec = [1 0 0 0 0 0 0 0 0 0 0 0];
49  nrcsw = [1 3 4 5 7 8 9 11];
50  segment = [2 6 10]; % The fault location options.
51  fault_loc = randsrc(1,1,segment);
52  if fault_loc == 6
53      fprintf('The fault occurs in branch 1.\n');
54  elseif fault_loc == 10
55      fprintf('The fault occurs in branch 2.\n');
56  else
57      fprintf('The fault occurs in before branches.\n');
58  end
```

```
59   end
60
61
62   %-------------------------------------------------------------------%
63   % Case 3: 15-nodes & 14-edges with 2 branches example.
64   %-------------------------------------------------------------------%
65   if casenumb == 3
66
67   IncM = [1 0 0 0 0 0 0 0 0 0 0 0 0 0
68          1 1 0 0 0 0 0 0 0 0 0 0 0 0
69          0 1 1 0 0 0 0 0 0 0 0 0 0 0
70          0 0 1 1 0 0 0 1 0 0 0 1 0 0
71          0 0 0 1 1 0 0 0 0 0 0 0 0 0
72          0 0 0 0 1 1 0 0 0 0 0 0 0 0
73          0 0 0 0 0 1 1 0 0 0 0 0 0 0
74          0 0 0 0 0 0 1 0 0 0 0 0 0 0
75          0 0 0 0 0 0 0 1 1 0 0 0 0 0
76          0 0 0 0 0 0 0 0 1 1 0 0 0 0
77          0 0 0 0 0 0 0 0 0 1 1 0 0 0
78          0 0 0 0 0 0 0 0 0 0 1 0 0 0
79          0 0 0 0 0 0 0 0 0 0 0 1 1 0
80          0 0 0 0 0 0 0 0 0 0 0 0 1 1
81          0 0 0 0 0 0 0 0 0 0 0 0 0 1];
82
83   connect_sta =[1 1 1 1 1 1 1 1 1 1 1 1 1 1 1];
84   sourceVec = [1 0 0 0 0 0 0 0 0 0 0 0 0 0 0];
85   nrcsw = [1 3 4 5 7 8 9 11 12 14];
86   segment = [2 6 10 13]; % The fault location options.
87   fault_loc = randsrc(1,1,segment);
88   if fault_loc == 6
89       fprintf('The fault occurs in branch 1.\n');
90   elseif fault_loc == 10
91       fprintf('The fault occurs in branch 2.\n');
92   elseif fault_loc == 13
93       fprintf('The fault occurs in branch 3.\n');
94   else
95       fprintf('The fault occurs in before branches.\n');
96   end
97   end
```

Listing 8.2
The Crew Coordination Algorithm

```
1    % Initialize output vectors.
2    engz_change = zeros(1,length(IncM(:,1)));
3    connect_change = ones(1,length(IncM(1,:)));
4
5    % The fault occurs randomly.
6    connect_sta(fault_loc) = 0;
7
8    % Input which nrcsw is close to the crew.
9    crew_loc = input('which nrcsw is close to the crew: ');
10   connect_sta(nrcsw(crew_loc)) = 0; % open the nrcsw.
11
12   % Energization of the initial state.
```

```matlab
13   engz_vec = ntp(IncM, connect_sta, sourceVec); % determine ...
         energization state.
14   head_nrcsw = find(IncM(:,nrcsw(crew_loc))==1,1,'first'); % ...
         find the head node of the opened nrcsw.
15   head_nrcsw_sit1 = head_nrcsw;
16
17   % Search direction from head to tail.
18   while sum(find(engz_vec(1:head_nrcsw_sit1)==0)) == 0 % if the ...
         nodes before the opened nrcsw are all energized.
19       for i = (crew_loc+1):length(nrcsw) % search from feeder ...
             head to tail.
20           connect_sta(nrcsw(i-1)) = 1; % close the privious nrcsw.
21           connect_sta(nrcsw(i)) = 0; % open the next nrcsw.
22           engz_vec = ntp(IncM, connect_sta, sourceVec);
23           head_nrcsw_sit1 = find(IncM(:,nrcsw(i))==1,1,'first');
24           connect_change = [connect_change;connect_sta];
25           engz_change = [engz_change;engz_vec];
26           if sum(find(engz_vec(1:head_nrcsw_sit1)==0)) ~= 0 % if ...
                 the fault segment appear, break the loop.
27               break
28           end
29       end
30       fprintf('the fault is between nrcsw %d and %d\n',i-1,i);
31   end
32
33   % Search direction from tail to head.
34   while sum(find(engz_vec(1:head_nrcsw)==0)) ~= 0 % if the nodes ...
         before the opened nrcsw are not all energized.
35       for i = (crew_loc-1):-1:1 % search from feeder tail to head.
36           connect_sta(nrcsw(i+1)) = 1;
37           connect_sta(nrcsw(i)) = 0;
38           engz_vec = ntp(IncM, connect_sta, sourceVec);
39           head_nrcsw = find(IncM(:,nrcsw(i))==1,1,'first');
40           connect_change = [connect_change;connect_sta];
41           engz_change = [engz_change;engz_vec];
42           if sum(find(engz_vec(1:head_nrcsw)==0)) == 0 % if the ...
                 search passed the fault segment, break the loop.
43               break
44           end
45       end
46       fprintf('the fault is between nrcsw %d and %d\n',i,i+1);
47   end
48
49   connect_change
50   engz_change
```

8.3   Scenarios for Outage Management

In this section, nine scenarios will be discussed for the outage management
system. The first four scenarios are without smart meters, while the latter
four cases have smart meters. The last scenario shows the switching proce-

dure between two feeders. The preliminary observation is that the last four scenarios really do not have too much impact on reliability because sometimes the trouble call tickets may help crews to identify the fault quickly. From the utility's perspective, revenue depends on how much reliable service they can provide to customers. For the situations in which the fault occurs in the primary network, whether or not there are smart meters has limited impact on the fault searching process.

Scenarios 1 (without smart meters) and 5 (with smart meters)

As shown in Fig. 8.7, the feeder is 10 miles long but no normally open tie switches connect with an adjacent feeder or other substation. The breaker in this scenario is tripped since a fault occurs in the middle of this feeder. Because there are no automatic switches or remote-controlled switches with fault indicators on the primary network, the crew will search for everything along the 10-miles-long feeder. The search time could be hours to a day.

Scenario 5 has the same conditions as Scenario 1 but with smart meters in the secondary network. However, the fault identification process is mostly the same as the case without smart meters since the fault indicator is usually installed through the line segment with remote-controlled switches in the primary network. The function of smart meters in this scenario is like the trouble call tickets in Scenario 1.

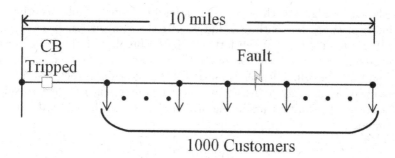

FIGURE 8.7
Schematics of Scenarios 1 and 5.

Scenarios 2 (without smart meters) and 6 (with smart meters)

The schematic of Scenarios 2 and 6 is illustrated in Fig. 8.8. In Scenario 2 the fault occurs on the secondary network. There are 1000 customers in this feeder. Four of the 1000 customers are de-energized because the fuse is burned. The customers' trouble call tickets will help the crew to localize the fault as well as the smart meters in Scenario 6. The smart meters may reduce the searching time from 2 hours to 1 hour since the mechanical reaction should be faster than the artificial reaction. However, the utility should balance the benefit of installing the smart meters to all customers and the revenue losses during a secondary network fault.

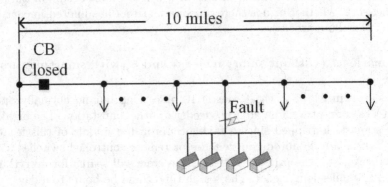

FIGURE 8.8
Schematics of Scenarios 2 and 6.

Scenarios 3 (without smart meters) and 7 (with smart meters)

As shown in Fig. 8.9, these two scenarios illustrate a fault occuring at the primary network but with an additional recloser to sectionalize the long feeder. If the fault occurs in the latter half, the breaker in the feeder head will not be tripped. If the fault occurs in the former part, the breaker will be tripped so that the whole feeder is affected. The searching processes of fault location in situations with and without smart meters might not display much difference since the fault occurs in the primary network. The affected number of customers in these two scenarios is 500.

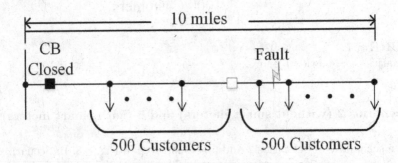

FIGURE 8.9
Schematics of Scenarios 3 and 7.

Scenarios 4 (without smart meters) and 8 (with smart meters)

Scenarios 4 and 8 as shown in Fig. 8.10 have SCADA capability to get data from the pole-mounted devices (FRTU/RTU) to feedback the information response from the fault indicators. Then the people in the control center will be able to analyze the entire network. The number of affected customers, in this case, will be 1000 at first. When the fault is localized and the automatic switch with fault indicator shown as "Yes" is tripped and the breaker is closed to restore the service for the healthy part, the number of de-energized customers will be 800. The role of smart meters is not obvious in this situation.

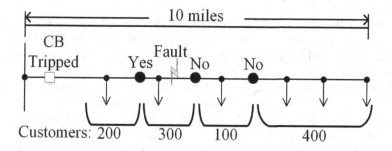

FIGURE 8.10
Schematics of Scenarios 4 and 8.

Scenario 9

Scenario 9 shows a feeder connected with an adjacent feeder or other substation by a normally open tie switch, as shown in Fig. 8.11. In addition, there are lots of non-remote-controlled switches (gray circles) along the line segment. When a fault occurs in the line segment, the breaker or recloser is tripped first so that the number of affected customers will be 1000. After the process of fault localization, the healthy parts will be temporarily restored so that the number of de-energized customers is 300. Following the switching procedure based on the crew coordination in the fault segment, the number of affected customers could go down iteratively during the temporary arrangement.

8.4 Energization States of a Subsystem

In the data exchange between the control center and the DMS, the operation status information will be sent to the physical system to change the

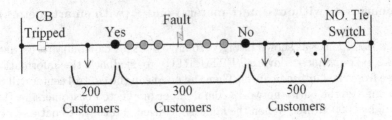

FIGURE 8.11
Schematics of Scenario 9.

state of the subsystems according to system requirements. The switch status would change the topology in a real-world situation. In order to achieve the monitor and control purposes during operations, two states, energization and de-energization, can be represented as 1 and 0, respectively.

The enumerated cases are the strategic infrastructure enhancement with 5 major milestones. Case 1 reflects the most common distribution system today. Cases 2 and 3 are the ongoing enhancement, while Cases 4 and 5 are networked microgrids. The element "E" represents "emergency" and "XE" is "extreme emergency." The energization states analysis does not focus on a single fault in a single feeder, but massive disturbances on multiple feeders. An example is when a tornado goes through a certain part of a distribution network. The energization analysis is based on the topology update given by the change of the switch states to estimate how many customers would be affected.

Case 1: Without DMS (E and XE)

Case 2: DMS system without tie-switches (E and XE)

Case 3: DMS with tie switches (E and XE)

Case 4: DMS system with NO tie-switches + DG + RCS (E and XE)

Case 5: DMS system with tie-switches + DG + RCS + microgrids (E and XE)

Algorithm 6 is the pseudocode of the energization status computing algorithm. The corresponding descriptions of relative elements in this algorithm have been enumerated below.

M_i Incidence matrix representing system topology. If vertex i is connected to edge j, then $M_i[i, j] = 1$, otherwise, 0.

M_a Adjacent matrix representing system topology. If vertex i is adjacent to vertex j, then $M_a[i, j] = 1$, otherwise, 0.

V_r Row vector indicating the open/close of switches/breakers/reclosers. If switch i is closed, then $V_r[i] = 1$, otherwise, 0.

Algorithm 6 Energization Status Computing Algorithm

Require: $\mathbf{M_i}, \mathbf{V_r}, \mathbf{V_s}, \mathbf{C}$

Ensure: energization status $\mathbf{V_e}$, number of total affected customers $\mathbf{N_T}$

$\mathbf{M_i} \leftarrow \mathbf{M_i} \times \mathbf{diag}(\mathbf{V_r})$;

$\mathbf{M_a} \leftarrow \mathbf{M_i} \times \mathbf{M_i^\top}$;

Replace all diagonal elements of $\mathbf{M_a}$ with 0; %Convert incidence to adjacency.

$\mathbf{V_e} \leftarrow \mathbf{0}$; %Initialize the number of energized vertices are 0s.

$\mathbf{V_e} \leftarrow \mathbf{V_s}$; %The source vertices are energized.

$\mathbf{V_e} \leftarrow \mathbf{V_e} \times \mathbf{M_a}$; %Find the energized vertices connected with source vertices.

$\mathbf{V_e} \leftarrow \mathbf{V_e} + \mathbf{V_s}$; %Add the source vertices.

Replace all non-zero elements in $\mathbf{V_e}$ with 1: %There may be elements counted more than once.

$\mathbf{N_T} \leftarrow (\mathbf{1} - \mathbf{V_e}) \times \mathbf{C^\top}$;

return $\mathbf{V_e}, \mathbf{N_T}$

V_s Row vector indicating the location of power sources. If vertex i is a substation-transformer or associated with a distribution generation (DG), then $V_s[i] = 1$, otherwise, 0.

V_e Row vector indicating energization status. If vertex i is energized, then $V_e[i] = 1$, otherwise, 0.

C Row vector indicating the number of customers associated with each subsystem node.

The MATLAB programming script of energization status computing is demonstrated in Listing 8.3. The number of affected customers under each affected area and its de-energization status can be calculated according to Listing 8.4. An example is applied to demonstrate the processes in this analysis. Shown in Fig. 8.12 are two equivalent topologies with different configurations. This is a simple 4-node network with 3 edges, where those edges are switches. Nodes 1 and 3 are the power injection resources and the edge 3 represents a normally open (NO) switch. The corresponding incidence matrix of this topology and all input vectors are shown below.

Listing 8.3

Energization Status Computing Function Based on the Incidence Matrix

```
1  % Energization Status Computing Based on Incidence Matrix
2  % INPUT: IncM is the incidence matrix
3  % INPUT: onOffVec is the row vector indicating the open/closed of
4  %                 switches/breakers/reclosers.
5  % INPUT: sourceVec is the row vector indicating the location ...
       of power sources.
```

```
6  % INPUT: cust_vec is the row vector indicating the number of ...
       customers
7  %                    associated with each sub-system node.
8
9  function engz_vec = engst(IncM, onOffVec, sourceVec, cust_vec)
10
11 % update the topology of the system
12 onOffMat = diag(onOffVec);
13 IncM = IncM * onOffMat;
14
15 % convert to adjacency matrix
16 adjMat = IncM * IncM';
17 adjMat = abs(adjMat - diag(diag(adjMat)));
18
19 engz_vec = sourceVec;
20 engz_old = engz_vec*0;
21
22 while (length(find(engz_old == 0)) - length(find (engz_vec == ...
       0))) ~= 0
23     engz_old = engz_vec;
24     engz_vec = engz_old * adjMat + engz_old;
25     idx = find(engz_vec > 0);
26     engz_vec(idx) = 1;
27
28 end
29 Ncust=(1-engz_vec)*cust_vec';
```

Listing 8.4
Function to Calculate the Number of Affected Customers

```
1  function Ncust = affcustomer(incMat, onOffVec, sourceVec,cust_vec)
2
3  %[x,y] = size(incMat);
4  %onOffVec = ones(1,y);
5  onOffMat = diag(onOffVec);
6  %onOffMat(7,7) = 0;
7  incMat = incMat * onOffMat;
8  adjMat = incMat*incMat';
9  adjMat = abs(adjMat - diag(diag(adjMat)));
10
11 engz_vec = sourceVec;
12 %engz_vec = [1,0,0,0,1,0,0,0,1,0,0,0,1,0,0,0];
13 engz_old = engz_vec*0;
14
15 while (length(find(engz_old == 0)) - length(find (engz_vec == ...
       0))) ~= 0
16     engz_old = engz_vec;
17     engz_vec = engz_old * adjMat + engz_old;
18     idx = find(engz_vec > 0);
19     engz_vec(idx) = 1;
20
21 end
22 Ncust=(1-engz_vec)*cust_vec';
```

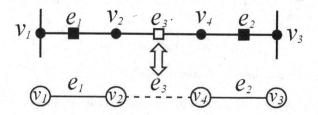

FIGURE 8.12
Example with 4-node system energization state.

$$M_i = \begin{array}{c} \\ v_1 \\ v_2 \\ v_3 \\ v_4 \end{array} \begin{array}{ccc} e_1 & e_2 & e_3 \\ \left(\begin{array}{ccc} 1 & 0 & 0 \\ 1 & 0 & 1 \\ 0 & 1 & 0 \\ 0 & 1 & 1 \end{array} \right) \end{array}, \ V_r = [1 \ 0 \ 1], \ V_s = [1 \ 0 \ 1 \ 0], \ C = [0 \ 3000 \ 0 \ 3000]$$

The input matrix of Algorithm 6 is the incidence matrix rather than the adjacency matrix because the incidence matrix can display both the vertices (columns) and edges (rows) directly. The row vector indicating the switches' states can update the connection status in the incidence matrix straightaway. However, the first step of Algorithm 6 is to convert the incidence matrix to an adjacency matrix, since the energization states are reflected on nodes but not edges. The conversion from the incidence matrix to the adjacency matrix is shown as:

$$M_a = M_{i(4\times3)} \times M_{i(3\times4)}^{\mathsf{T}} = \begin{array}{c} \\ v_1 \\ v_2 \\ v_3 \\ v_4 \end{array} \begin{array}{cccc} v_1 & v_2 & v_3 & v_4 \\ \left(\begin{array}{cccc} 1 & 1 & 0 & 0 \\ 1 & 2 & 0 & 1 \\ 0 & 0 & 1 & 1 \\ 0 & 1 & 1 & 2 \end{array} \right) \end{array}.$$

Replace all the diagonal elements with zeros to obtain the intended adjacency matrix:

$$\widehat{M}_a = \begin{array}{c} \\ v_1 \\ v_2 \\ v_3 \\ v_4 \end{array} \begin{array}{cccc} v_1 & v_2 & v_3 & v_4 \\ \left(\begin{array}{cccc} 0 & 1 & 0 & 0 \\ 1 & 0 & 0 & 1 \\ 0 & 0 & 0 & 1 \\ 0 & 1 & 1 & 0 \end{array} \right) \end{array}.$$

Then, update the V_e:

$$V_e = V_{s(1\times4)} \times \widehat{M}_{a(4\times4)} = [1 \ 1 \ 0 \ 1];$$

The updated V_e is then incorporated with the vertices that connected with the injection sources V_s:

$$V_e' = V_e + V_s = [2\ 1\ 1\ 1];$$

Convert all non-zero constant elements into 1s:

$V_e'' = [1\ 1\ 1\ 1] \to V_e$, which is the row vector indicating energization status.

Finally, the total number of affected customers under this scenario is $N_T = [1 - V_e] \cdot C^{\mathsf{T}} = 0$, where $[1 - V_e]$ determine the energization state of the subsystem.

When a fault is assumed to occur on e_2 and the tie switch on e_2 is opened, the energization states vector can be updated according to the previous steps, and obtained as:

$$V_e = [1\ 1\ 1\ 0].$$

Calculate the total number of affected customers:

$$N_T = [1 - V_e] \cdot C^{\mathsf{T}} = 3{,}000 \text{ affected customers.}$$

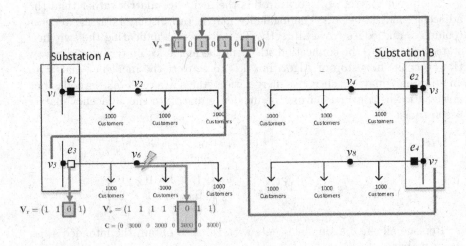

FIGURE 8.13
Example with 2-Substations energization state.

Another example as shown in Fig. 8.13 is the emergency situation in Case 1, of distribution feeders without DMS. Since all switches in the extreme emergency situation would be breaked, we only consider the emergency situation. There are 2 substations and 4 feeders in this network and the equivalent graph representation consists of 8 vertices and 4 edges. The corresponding incidence matrix of this network is shown below, while the other input vectors V_r, V_s, C, and the energization result vector V_e can be obtained according to the topology and the switches' states. The estimated number of affected customers in this case is 3,000.

$$
M_i = \begin{array}{c} \\ v_1 \\ v_2 \\ v_3 \\ v_4 \\ v_5 \\ v_6 \\ v_7 \\ v_8 \end{array}
\begin{array}{cccc}
e_1 & e_2 & e_3 & e_4 \\
\left(\begin{array}{cccc}
1 & 0 & 0 & 0 \\
1 & 0 & 0 & 0 \\
0 & 1 & 0 & 0 \\
0 & 1 & 0 & 0 \\
0 & 0 & 1 & 0 \\
0 & 0 & 1 & 0 \\
0 & 0 & 0 & 1 \\
0 & 0 & 0 & 1
\end{array}\right)
\end{array}.
$$

8.5 Evaluation of Reliability Indices

People consume more electricity than ever before and expect to get it without interruptions. Authorities are increasingly interested in how well electricity is provided to people. This puts pressure on electrical companies to provide evidence of the reliability of their electricity supply and distribution. Today, it is fairly common for utilities to measure the time duration, the number of out-of-service customers, and the number of interruptions in the distribution system. For this they use international standards called system average interruption duration index (SAIDI), system average interruption frequency index (SAIFI), and customer average interruption duration index (CAIDI), created by the Institute of Electrical Electronics Engineers (IEEE) [181].

SAIDI is a system index of the average duration of an interruption in the power supply indicated in minutes per customer. It is used as a performance measure to compare switching strategies [181]. SAIDI is defined as:

$$
\text{SAIDI} = \frac{\text{Sum of Customer Interruption Durations}}{\text{Number of Customers Served}}.
$$

SAIDI is used as a performance measure, since restoration strategies primarily impact the annual interruption time experienced by customers. Additionally, SAIDI is one of the most common reliability indices used by distribution utilities, and it is easily computed from individual customer results, available from an analytical simulation [152].

SAIFI is another system index of average frequency of interruptions in power supply. It is defined as:

$$
\text{SAIFI} = \frac{\text{Total Number of Customers' Interruptions}}{\text{Number of Customers Served}}.
$$

SAIFI is improved by reducing the frequency of outages and by reducing the number of customers interrupted when outages do occur.

SAIFI measures how often a feeder or a system could cause a power outage such that the customers within this area can expect to experience an outage. SAIDI measures the average outage duration per customer under a feeder or system, and CAIDI measures average outage duration within the system if an outage is experienced, or average restoration time [181].

CAIDI is defined as:

$$\text{CAIDI} = \frac{\text{Sum of Customer Interruption Durations}}{\text{Total Number of Customers' Interruptions}},$$

which is similar to:

$$\text{CAIDI} = \frac{\text{SAIDI}}{\text{SAIFI}}.$$

Even though CAIDI is not a straightforward index to represent the quality of service, it is affected by SAIDI and SAIFI. Since SAIFI and SAIDI are driven primarily by the outage frequency or the time duration, respectively, both variables can adversely affect CAIDI.

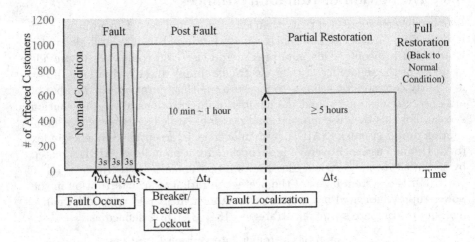

FIGURE 8.14
Change in the number of affected customers from a fault-to-full restoration (the exact number of affected customers on each phase depends on feeder topology).

Fig. 8.14 demonstrates the change in the number of affected customers during four operational states: (1) a detection of permanent fault, (2) post fault period, (3) isolate fault for minimal interruption and partial restoration, and 4) full restoration. Once a fault occurs, and the tie switch on the feeder head is tripped, all customers on this feeder (in Fig. 8.14 we assume there are 1,000 customers for a given feeder) are affected by the power outage. The recloser (remote-controlled switch) attempts to restore the connection every 3 seconds ($\Delta t_{1,2,3}$). After 3 attempts, the recloser is locked out where the faulted area experiences a significant power outage. Then, the fault is localized and isolation of the fault segregates the segment from other healthy parts of the feeder in order to narrow down the faulted area and achieve partial restoration. This process usually takes 10 minutes to 1 hour at most. After at least 5 hours of repair duration, the operator recloses the tie switch and the system is back to the normal condition (full restoration to the original state).

In order to show how to calculate SAIDI, SAIFI, and CAIDI according to the system topology, the 9 scenarios discussed in the previous section will be utilized to demonstrate the results with different characteristics. As mentioned, Scenarios 1 to 4 are without smart meters, while Scenarios 5 to 9 are with smart meters. The fault response time, which is a part of the power outage time, will be shorter if customers are equipped with smart meters. All cases have fuses and breakers. Cases 3, 4, 7, 8, and 9 have remote-controlled switches. Only the feeder in Scenario 9 connects with the adjacent feeder or substation through the normally open tie switch. Before performing the calculation, something more concrete about the outage time and affected customers should be defined. The failure rate, which is the number of failure times per year and the total outage time in hours, can be assumed and obtained from the former discussion. The number of affected customers can be estimated from the energization state analysis. Therefore, SAIDI and SAIFI for a system per year can be defined as:

$$\text{SAIDI/year/syst} = \frac{N_T \cdot t_{\text{outage}}}{\sum C_{\text{syst}}} = \frac{[1 - V_e] \cdot C^\top \cdot t_{\text{outage}}}{\sum C_{\text{syst}}},$$

$$\text{SAIFI/year/syst} = \frac{N_T \cdot f}{\sum C_{\text{syst}}} = \frac{[1 - V_e] \cdot C^\top \cdot f}{\sum C_{\text{syst}}}, \forall f = 1, 2, \cdots, F,$$

where N_T is the total number of affected customers, Time indicates the interruption duration, C_{syst} is a row vector indicating the number of customers associated with the system, V_e is the row vector indicating energization status, f is the failure rate of the system, and t_{outage} is the total sum-up outage time, i.e., $t_{\text{outage}} = \Delta t_{1-5}$. The corresponding values of SAIDI, SAIFI, and CAIDI of the 9 scenarios are set in Fig. 8.15.

	Without Smart Meters				With Smart Meters				
	1	2	3	4	5	6	7	8	9
With Fuses	x	x	x	x	x	x	x	x	x
With Breaker	x	x	x	x	x	x	x	x	x
With Recloser			x	x			x	x	x
With Remote-Controlled switches				x				x	x
With normally open (NO) Switches									x
Failure Rate (times/year)	1	1	1	1	1	1	1	2	1
Outage time (hours)	5	2	5	5 minutes then 4 hours	5	1	4	5 minutes then 4 hours	1 minute, then 5 minutes, then 3 hours
Affected Customers	1000	4	500	1000, then 800 customers	1000	4	500	1000, then 800 customers	1000, then 300, then … could go down to 20 customers during temporary arrangement
System Average Interruption Duration Index (SAIDI)	5.000	0.008	2.500	3.283	5.000	0.004	2.000	3.283	0.122
System Average Interruption Frequency Index (SAIFI)	1.000	0.004	0.500	1.000	1.000	0.004	0.500	2.000	1.000
Customer Average Interruption Duration Index (CAIDI)	5.000	2.000	5.000	3.283	5.000	1.000	4.000	1.642	0.122

FIGURE 8.15

Summary of 9 scenarios and detailed number of SAIDI, SAIFI, and CAIDI associated with each feeder/subsystem.

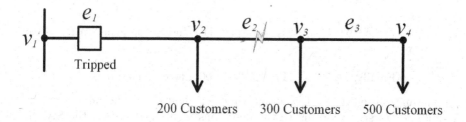

FIGURE 8.16
Schematic of Scenario 1.

Fig. 8.16 is the topological representation of Scenario 1. Once a fault occurs in this system, the tie switch on the feeder head will be tripped so that the whole system will be de-energized. For energization state analysis, the input incidence matrix and all relative vectors of this system are shown as:

$$M_i = \begin{array}{c} \\ v_1 \\ v_2 \\ v_3 \\ v_4 \end{array}\begin{array}{c} e_1 \quad e_2 \quad e_3 \\ \begin{pmatrix} 1 & 0 & 0 \\ 1 & 1 & 0 \\ 0 & 1 & 1 \\ 0 & 0 & 1 \end{pmatrix} \end{array}, \quad V_r = [0\ 1\ 1], \quad V_s = [1\ 0\ 0\ 0], \quad C = [0\ 200\ 300\ 500].$$

Updating the topology with V_r,

$$M_i = M_i \times \text{Diag}(V_r),$$

the conversion from the incidence matrix to the adjacency matrix is shown as:

$$M_a = M_{i(4\times3)} \times M_{i(3\times4)}^\mathsf{T} = \begin{array}{c} \\ v_1 \\ v_2 \\ v_3 \\ v_4 \end{array}\begin{array}{c} v_1 \quad v_2 \quad v_3 \quad v_4 \\ \begin{pmatrix} 0 & 0 & 0 & 0 \\ 0 & 1 & 1 & 0 \\ 0 & 1 & 2 & 1 \\ 0 & 0 & 1 & 1 \end{pmatrix} \end{array}.$$

Replace all the diagonal elements with zeros to obtain the intended adjacency matrix:

$$\widehat{M_a} = \begin{array}{c} \\ v_1 \\ v_2 \\ v_3 \\ v_4 \end{array}\begin{array}{c} v_1 \quad v_2 \quad v_3 \quad v_4 \\ \begin{pmatrix} 0 & 0 & 0 & 0 \\ 0 & 0 & 1 & 0 \\ 0 & 1 & 0 & 1 \\ 0 & 0 & 1 & 0 \end{pmatrix} \end{array}.$$

Then, calculate:

$$V_e = V_{s(1\times4)} \times \widehat{M}_{a(4\times4)} = [0\ 0\ 0\ 0];$$

$V_e' = V_e + V_s = [1\,0\,0\,0]$, which is the row vector indicating energization status.

Finally, $N_T = [1 - V_e] \cdot C^\top = 1000$ affected customers.

Assume the entire feeder outage time is 5 hours and the failure rate is 1 time/year/syst. According to the equations mentioned before, the SAIDI, SAIFI, and CAIDI can be calculated as:

$$\text{SAIDI} = \frac{1000 \text{ customers} \times 5 \text{ hours}}{1000 \text{ customers}} = 5,$$

$$\text{SAIFI} = \frac{1000 \text{ customers} \times 1 \text{ times/year}}{1000 \text{ customers}} = 1,$$

$$\text{CAIDI} = \frac{\text{SAIDI}}{\text{SAIFI}} = 5.$$

The value of 5 of SAIDI and CAIDI represents low reliability in the one feeder system.

Similarly, to calculate the indexes in Scenario 2,

$$\text{SAIDI} = \frac{4 \text{ customers} \times 2 \text{ hours}}{1000 \text{ customers}} = 0.008,$$

$$\text{SAIFI} = \frac{4 \text{ customers} \times 1 \text{ times/year}}{1000 \text{ customers}} = 0.004,$$

$$\text{CAIDI} = \frac{\text{SAIDI}}{\text{SAIFI}} = 2.$$

Fig. 8.17 is the topological representation of Scenario 8. Once a fault occurs in this system, the tie switch on the feeder head will be tripped so that the whole system will be de-energized. Then, the remote-controlled switch (RCS) can help to narrow down and isolate the faulted area. For energization state analysis, the input incidence matrix and all relative vectors of this system are shown as:

$$M_i = \begin{array}{c} \\ v_1 \\ v_2 \\ v_3 \\ v_4 \\ v_5 \end{array} \begin{array}{cccc} e_1 & e_2 & e_3 & e_4 \\ \left(\begin{array}{cccc} 1 & 0 & 0 & 0 \\ 1 & 1 & 0 & 0 \\ 0 & 1 & 1 & 0 \\ 0 & 0 & 1 & 1 \\ 0 & 0 & 0 & 1 \end{array} \right) \end{array}, \quad V_r = [0\,1\,1\,1], \quad V_s = [1\,0\,0\,0\,0],$$

$$C = [0\ 200\ 300\ 100\ 400].$$

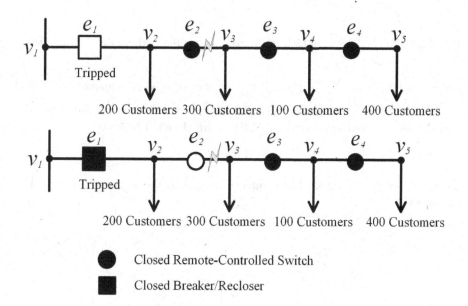

FIGURE 8.17
Feeder statuses for Scenario 8.

Update the topology with V_r; the conversion from the incidence matrix to the adjacency matrix is shown as:

$$M_a = M_{i(5\times4)} \times M_{i(4\times5)}^T = \begin{array}{c} \\ v_1 \\ v_2 \\ v_3 \\ v_4 \\ v_5 \end{array} \begin{pmatrix} \begin{array}{ccccc} v_1 & v_2 & v_3 & v_4 & v_5 \\ 0 & 0 & 0 & 0 & 0 \\ 0 & 1 & 1 & 0 & 0 \\ 0 & 1 & 2 & 1 & 0 \\ 0 & 0 & 1 & 2 & 1 \\ 0 & 0 & 0 & 1 & 1 \end{array} \end{pmatrix}.$$

Replace all the diagonal elements with zeros to obtain the intended adjacency matrix:

$$\widehat{M}_a = \begin{array}{c} \\ v_1 \\ v_2 \\ v_3 \\ v_4 \\ v_5 \end{array} \begin{pmatrix} \begin{array}{ccccc} v_1 & v_2 & v_3 & v_4 & v_5 \\ 0 & 0 & 0 & 0 & 0 \\ 0 & 0 & 1 & 0 & 0 \\ 0 & 1 & 0 & 1 & 0 \\ 0 & 0 & 1 & 0 & 1 \\ 0 & 0 & 0 & 1 & 0 \end{array} \end{pmatrix}.$$

Then, calculate:

$$V_e = V_{s(1\times4)} \times \widehat{M}_{a(4\times4)} = [0\ 0\ 0\ 0\ 0];$$

$V_e' = V_e + V_s = [1\,0\,0\,0\,0]$, which is the row vector indicating energization status.

Therefore, $N_T = [1 - V_e] \cdot C^\top = 1000$ affected customers.

After the fault localization and isolation, the tie switch at the feeder head and the remote-controlled switch (RCS) e_2 are closed. The vector

$$V_r = [1\,1\,0\,0].$$

Then, the conversion from the incidence matrix to the adjacency matrix is shown as:

$$M_a = M_{i(5\times4)} \times M_{i(4\times5)}^\top = \begin{array}{c} \\ v_1 \\ v_2 \\ v_3 \\ v_4 \\ v_5 \end{array} \begin{array}{c} \begin{array}{ccccc} v_1 & v_2 & v_3 & v_4 & v_5 \end{array} \\ \left(\begin{array}{ccccc} 1 & 1 & 0 & 0 & 0 \\ 1 & 2 & 1 & 0 & 0 \\ 0 & 1 & 1 & 0 & 0 \\ 0 & 0 & 0 & 0 & 0 \\ 0 & 0 & 0 & 0 & 0 \end{array} \right) \end{array}.$$

Replace all the diagonal elements with zeros to obtain the intended adjacency matrix:

$$\widehat{M_a} = \begin{array}{c} \\ v_1 \\ v_2 \\ v_3 \\ v_4 \\ v_5 \end{array} \begin{array}{c} \begin{array}{ccccc} v_1 & v_2 & v_3 & v_4 & v_5 \end{array} \\ \left(\begin{array}{ccccc} 0 & 1 & 1 & 0 & 0 \\ 1 & 0 & 1 & 0 & 0 \\ 0 & 1 & 0 & 0 & 0 \\ 0 & 0 & 0 & 0 & 0 \\ 0 & 0 & 0 & 0 & 0 \end{array} \right) \end{array}.$$

Calculate:

$$V_e = V_{s(1\times4)} \times \widehat{M}_{a(4\times4)} = [0\,1\,0\,0\,0];$$

$V_e' = V_e + V_s = [1\,1\,0\,0\,0]$, which is the row vector indicating energization status.

Therefore, $N_T = [1 - V_e] \cdot C^\top = 800$ affected customers.

Assume the fault isolation time will be 5 minutes, the entire feeder outage time is 4 hours, and the failure rate is 2 times/year/syst. According to the equations mentioned before, the SAIDI, SAIFI, and CAIDI can be calculated as:

$$\text{SAIDI} = \frac{1000 \text{ customers} \times 5/60 \text{ hours} + 800 \text{ customers} \times 4 \text{ hours}}{1000 \text{ customers}} = 3.283,$$

$$\text{SAIFI} = \frac{1000 \text{ customers} \times 2 \text{ times/year}}{1000 \text{ customers}} = 2,$$

$$\text{CAIDI} = \frac{\text{SAIDI}}{\text{SAIFI}} = 1.642.$$

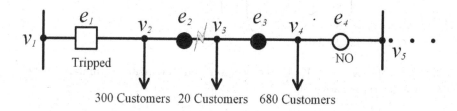

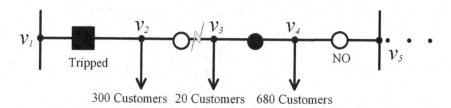

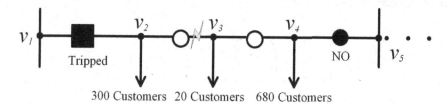

FIGURE 8.18
Feeder statuses for Scenario 9.

In Scenario 9, because there are remote-controlled switches (RCS) along the feeder and the feeder is connected to a redundant power injection source, the outage time could be 1 minute for the whole feeder (1000 customers are affected), 5 minutes to isolate the fault area (300 customers are affected), and the switching procedure will narrow down the faulted section to 20 customers to go through the 3 hours' outage time. As shown in Fig. 8.18, the input incidence matrix and all relative vectors of this system for the energization state analysis are shown as:

$$M_i = \begin{array}{c} \\ v_1 \\ v_2 \\ v_3 \\ v_4 \\ v_5 \end{array} \begin{array}{c} e_1 \; e_2 \; e_3 \; e_4 \\ \begin{pmatrix} 1 & 0 & 0 & 0 \\ 1 & 1 & 0 & 0 \\ 0 & 1 & 1 & 0 \\ 0 & 0 & 1 & 1 \\ 0 & 0 & 0 & 1 \end{pmatrix} \end{array}, \; V_r = [0\,1\,1\,0], \; V_s = [1\,0\,0\,0\,1],$$

$$C = [0\ 700\ 20\ 280\ 0].$$

Update the topology with V_r; the conversion from the incidence matrix to the adjacency matrix is shown as:

$$M_a = M_{i(5\times4)} \times M_{i(4\times5)}^\mathsf{T} = \begin{array}{c} \\ v_1 \\ v_2 \\ v_3 \\ v_4 \\ v_5 \end{array} \begin{array}{c} v_1 \; v_2 \; v_3 \; v_4 \; v_5 \\ \begin{pmatrix} 0 & 0 & 0 & 0 & 0 \\ 0 & 1 & 1 & 0 & 0 \\ 0 & 1 & 2 & 1 & 0 \\ 0 & 0 & 1 & 2 & 1 \\ 0 & 0 & 0 & 0 & 0 \end{pmatrix} \end{array}.$$

Replace all the diagonal elements with zeros to obtain the intended adjacency matrix:

$$\widehat{M_a} = \begin{array}{c} \\ v_1 \\ v_2 \\ v_3 \\ v_4 \\ v_5 \end{array} \begin{array}{c} v_1 \; v_2 \; v_3 \; v_4 \; v_5 \\ \begin{pmatrix} 0 & 0 & 0 & 0 & 0 \\ 0 & 0 & 1 & 0 & 0 \\ 0 & 1 & 0 & 1 & 0 \\ 0 & 0 & 1 & 0 & 1 \\ 0 & 0 & 0 & 0 & 0 \end{pmatrix} \end{array}.$$

Then, calculate:

$$V_e = V_{s(1\times4)} \times \widehat{M}_{a(4\times4)} = [0\ 0\ 0\ 0\ 0];$$

$V_e' = V_e + V_s = [1\,0\,0\,0\,1]$, which is the row vector indicating energization status.

Therefore, $N_T = [1 - V_e] \cdot C^\mathsf{T} = 1000$ affected customers.

After the fault localization and isolation, the tie switch at the feeder head and the remote-controlled switch (RCS) e_2 are closed. The updated

$$V_r = [1\ 0\ 0\ 0].$$

Then, the conversion from the incidence matrix to the adjacency matrix is shown as:

$$M_a = M_{i(5\times4)} \times M_{i(4\times5)}^\top = \begin{array}{c} \\ v_1 \\ v_2 \\ v_3 \\ v_4 \\ v_5 \end{array} \begin{array}{ccccc} v_1 & v_2 & v_3 & v_4 & v_5 \\ \left(\begin{array}{ccccc} 1 & 1 & 0 & 0 & 0 \\ 1 & 1 & 0 & 0 & 0 \\ 0 & 0 & 0 & 0 & 0 \\ 0 & 0 & 0 & 0 & 0 \\ 0 & 0 & 0 & 0 & 0 \end{array}\right) \end{array}.$$

Replace all the diagonal elements with zeros to obtain the intended adjacency matrix:

$$\widehat{M_a} = \begin{array}{c} \\ v_1 \\ v_2 \\ v_3 \\ v_4 \\ v_5 \end{array} \begin{array}{ccccc} v_1 & v_2 & v_3 & v_4 & v_5 \\ \left(\begin{array}{ccccc} 0 & 1 & 0 & 0 & 0 \\ 1 & 0 & 0 & 0 & 0 \\ 0 & 0 & 0 & 0 & 0 \\ 0 & 0 & 0 & 0 & 0 \\ 0 & 0 & 0 & 0 & 0 \end{array}\right) \end{array}.$$

Calculate:

$$V_e = V_{s(1\times4)} \times \widehat{M_a}_{(4\times4)} = [0\ 1\ 0\ 0\ 0];$$

$V_e' = V_e + V_s = [1\ 1\ 0\ 0\ 1]$, which is the row vector indicating energization status.

Therefore, $N_T = [1 - V_e'] \cdot C^\top = 300$ affected customers.

During the duration of the repairing, the tie switch connect the faulted feeder with an adjacent feeder will close to minimize the number of affected customers. If this happens, the remote-controlled switches e_2 and e_3 will open to isolate the faulted area. Under this situation, V_r is updated as $[1\ 1\ 0\ 0\ 1]$ and the conversion from the incidence matrix to the adjacency matrix is shown as:

$$M_a = M_{i(5\times4)} \times M_{i(4\times5)}^\top = \begin{array}{c} \\ v_1 \\ v_2 \\ v_3 \\ v_4 \\ v_5 \end{array} \begin{array}{ccccc} v_1 & v_2 & v_3 & v_4 & v_5 \\ \left(\begin{array}{ccccc} 1 & 1 & 0 & 0 & 0 \\ 1 & 1 & 0 & 0 & 0 \\ 0 & 0 & 0 & 0 & 0 \\ 0 & 0 & 0 & 1 & 1 \\ 0 & 0 & 0 & 1 & 1 \end{array}\right) \end{array}.$$

Replace all the diagonal elements with zeros to obtain the intended adjacency matrix:

$$\widehat{M_a} = \begin{array}{c} \\ v_1 \\ v_2 \\ v_3 \\ v_4 \\ v_5 \end{array} \begin{array}{ccccc} v_1 & v_2 & v_3 & v_4 & v_5 \\ \left(\begin{array}{ccccc} 0 & 1 & 0 & 0 & 0 \\ 1 & 0 & 0 & 0 & 0 \\ 0 & 0 & 0 & 0 & 0 \\ 0 & 0 & 0 & 0 & 1 \\ 0 & 0 & 0 & 1 & 0 \end{array} \right) \end{array} .$$

Then, calculate:

$$V_e = V_{s(1 \times 4)} \times \widehat{M}_{a(4 \times 4)} = [0\ 1\ 0\ 1\ 0];$$

$V_e' = V_e + V_s = [1\ 1\ 0\ 1\ 1]$, which is the row vector indicating energization status.

Therefore, $N_T = [1 - V_e] \cdot C^\top = 20$ affected customers.

Assume the fault isolation time will be 1 minutes, the tie switch will close in 5 minutes, the repair duration is 3 hours, and the failure rate is 1 times/year/syst. According to the equations mentioned before, the SAIDI, SAIFI, and CAIDI can be calculated as:

$$\text{SAIDI} = \frac{(1000 \text{ cust.} \times 1/60 \text{ hours}) + (300 \text{ cust.} \times 5/60 \text{ hours}) + 3 \text{ hours}}{20 \text{ cust.}} = 0.122,$$

$$\text{SAIFI} = \frac{1000 \text{ cust.} \times 1 \text{ time/year}}{1000 \text{ cust.}} = 1,$$

$$\text{CAIDI} = \frac{\text{SAIDI}}{\text{SAIFI}} = 0.122.$$

Fig. 8.15 shows all scenarios with the detailed numbers for each scenario. Notice that the reliability can vary significantly when a technology has been deployed, to enable automation and facilitate the remedial actions with remote-controlled capability. As demonstrated in Scenarios 1, 8, and 9, similar calculation can be evaluated for the remaining, Cases, 2, 3, 4, 5, 6, and 7.

8.6 Conclusions

Outage management is the central core that coordinates a scheduled/unscheduled outage with an assignment of crew members to the site. It is also related to the coordination of switching steps that are safety-related. The sequential generation of switching orders based on the latest events helps to dispatch crew members and coordinate with the latest trouble calls, as well as the latest energization statuses from all switching devices. This chapter establishes interconnected feeders (with NO tie switches) to coordinate remote-controlled switches and non-remote-controlled switches to minimize the smallest outage (segment) where some non-remote-controlled switches require crew members at a site to coordinate with control center dispatchers. The latest topological state in a distribution network is established from the switch statuses that are connected with feeder remote terminal units (FRTUs). Hence, the graph is updated based on the latest FRTU information and determines the next reconfiguration to effectively isolate the smallest fault segment (between non-remote-controlled switches in a feeder) and partial restoration.

9

Switching Procedure

A power outage can be either unscheduled or scheduled. Chapters 5, 6, 7, and 8 focus on localizing a faulted segment of distribution feeder and temporary isolation and restoration for the repair period. Such an unscheduled outage is often very time consuming and unpredictable. Predicting the unpredictable and optimizing resources involves automatic generation of switching steps to handle the disturbance of an electrical fault occurrence.

In a scheduled outage, such similar reconfigurations with automatic generation of switching procedures can also be applied here for preventive maintenance. The major purpose of switching procedures is to schedule an outage for maintenance, which is often planned in advance with minimum interruption. The consideration of switching is often deterministic because of the specific location to be out of service. Therefore, the scheduling can be flexible based on the feeder condition. This chapter also includes the reconfiguration application in detecting inconsistent metering information between the sections of metered switches, i.e., RCSs, as well as its interdependence with another feeder(s). Reconfiguration without interrupting power to consumers is done to verify accounting information power flow for each feeder, to find anomalies such a potential fraud via tampering.

9.1 Scheduled Outage

The distribution network spans a large area in a region, from the feeder head where it is connected to a distribution substation and the rest of the feeder. Occasionally, ongoing maintenance may be necessary. However, before the work is scheduled to take the network offline, the crew may need to coordinate with the operators in the control room as to the latest conditions of a feeder section where it can be reconfigured without interruption. This will minimize the outage so that loads can be carried from another part of a network. As introduced in Chapter 2, the topology of a distribution network can be reconfigured if there are NO switches that are connected to the neighborhood of the network. The other motivation is that the systematic disconnect of a line segment can be verified in the control center, which provides a global view of what the feeder operating condition would be with "what if" power flow evaluation. Such scheduled outages can be optimally planned with the available crew schedules and their work hours, as well as the loading condition of a feeder at that time. Such planned out-of-services are typically made in advance, to extend the network, facilitate replacement of line, or devices, or ongoing preventive maintenance. Preventing service interruptions for customers is the priority. Lowering the cost of engineering and maintenance for conforming to regulations problems is always a crucial part of a scheduled outage.

The outage scheduling includes determination of work hours during the day and the time intervals for the outage in the service area. In addition, since the topology of the distribution system is usually radial, a scheduled outage provides the expected readiness to transfer loads uninterruptedly as compared to load in a faulted feeder, to avoid unnecessary interruption and reduce the number of affected customers. To keep the topology in radial configuration, problems can be modeled as discrete, nonlinear, and as having multiple objectives [182, 183]. Other associated constraints are the availability of crew members and associated switching operations for load transfer of downstream loads. As this scheduling problem can be limited to crew resources and their already assigned work hours, determining the optimal scheduled maintenance is a discrete chronological problem [184].

9.1.1   Scheduled and Unscheduled Outages

For an unscheduled outage, the crew is not scheduled and assigned a job, but the crew in a scheduled outage is responsible for a task in advance. An unscheduled outage is mainly because of a disturbance from a electrical fault. It can also be called a fault outage. Since the fault occurrence was not predictable, the power outage area is usually much larger than that of the scheduled outage. The coordination of the unscheduled outage management system is based on the fault indicator, the remote-controlled switches, and sometimes the smart meters, as well as the trouble call tickets from customers. For dealing with an unscheduled outage, the process requires coordinating with the crew member who is working on site to find the fault, and at the same time, the operator needs to temporarily rearrange the power from other adjacent feeders. Scheduled outages include planned maintenance outages and temporary construction outages. A scheduled outage is predictable and the crew knows the exact location of the segment or area of the temporary power outage. So, the scheduled outage node could be isolated ahead of time, and the operator could transfer the downstream fault-free load to the alternative interconnected feeder ahead of time.

Scheduled outages are usually affected by the complexity of the distribution grid. If the target area has more microgrids or clusters with multiple types of loads, the scheduled outage time will be much greater. The fault localizing and repair time of the component is also relevant. But an unscheduled outage is not affected by this [177].

9.1.2 Outage Scheduling Process

Electricity services interruption for planned work such as maintenance, facility replacement, or system expansion in distribution systems will involve arranging a scheduled outage. The flow of scheduling outage work in distribution system feeders is shown in Fig. 9.1. Three phases are cataloged in the procedure of scheduled outage engineering: preparation, scheduling, and implementation. In the preparation phase, data collection is done as the preliminary work require collecting the events logs, verifying the availability of resources, and determining the territories and duration of the scheduled events to be performed. Then an optimal-outage scheduling process can be executed according to the intelligent analysis, and the results should be verified. In the implementation phase, the customers in the planning areas should be notified by the utility before the power outage, and the down-stream load areas should keep their service during the work because of relative switching procedures. Besides, additional switching operations should be executed to restore the feeders after the work [184].

From practical considerations in intelligent outage scheduling, some guidelines are given [183]:

- One assignment for one engineering or planning team at one time (no more than two or more activities).

- Avoid unnecessary interruption(s).

- Plan the work within business hours except for areas serving special customers, who can pre-apply for a specific period of time.

- To minimize the impact of power outages (number of affected customers, outage duration, etc.), plan the work with minimal loading during the interruption.

- To make the scheduling more efficient, transportation distances of the engineering crews should be minimized.

- The time deviation among the same type of crews should be as small as possible.

9.1.3 Outage Scheduling Formulation

The probability of scheduled outages within distribution feeders is presumed to be similar. Statistically, the frequency of construction efforts for additional poles connecting to the existing feeder depends on the residential area of development in an area. The scheduled outage time of its feeder often has an estimated constant value as to the anticipated time for out-of-service events. Most scheduled outages involve incremental work that does not require de-

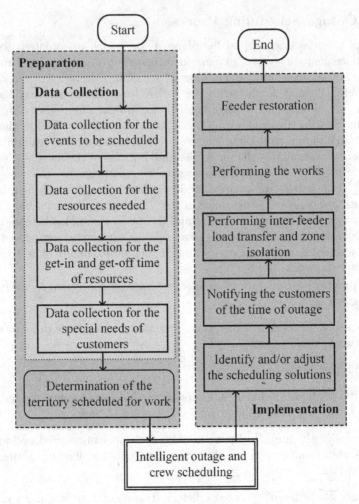

FIGURE 9.1
Procedure of scheduled outage engineering in distribution systems [184].

energizing a larger area. Therefore, each node has a scheduled outage frequency (SOF) that can be defined as:

$$\text{SOF} = \frac{\text{Scheduled outage frequency of a feeder}}{\text{The number of nodes in the feeder}}.$$

The outage scheduling for distribution feeders is the minimization of multiple calculating objectives under constraints that 1) avoid unnecessary service interruption, 2) avoid crew overtime, and 3) schedule work in specific zones on specific days and times. The problem can be formulated as follows [182]:

$$\text{Min } (L_O, T_W, T_S, W_D)$$

where

L_O is the overall outage loading of the scheduling,

T_W is the amount of working time of the scheduling,

T_S is the span of time for a feeder,

W_D is the divergence of crews or working teams for the zones in a feeder.

9.1.3.1 Minimize Outage Loading

The overall outage loading depends on the amount and types of loads in the service zones. The load pattern is diverse, since the seasonal consumption varies (summer and winter, mainly HVAC). In addition, the load types of residential, industrial, and commercial should be categorized in the analysis. The outage loading of the scheduling can be calculated by:

$$L_O = \sum_{z=1}^{Z} \left(R_z \cdot \sum_{i=1}^{N_{L,T}} \left(P_{z,i} \cdot \int_{t_{z,f}}^{t_{z,0}} f_{L,i}(t)dt \right) \right),$$

where

Z is the number of service zones to have scheduled outages,

R_z is the total outage rating of the service zone z,

$N_{L,T}$ is the number of typical load types, which include different seasons and categories,

$P_{z,i}$ is the percentage loading of the typical load of type i in zone z,

$t_{z,0}$ is the begin time of the outage for service zone z,

$t_{z,f}$ is the end time of the outage for service zone z,

$f_{L,i}(t)$ is the load profile function of typical load of type i.

The data of R_z, $N_{L,T}$, and $P_{z,i}$ can be calculated according to the data archived in the customer information system of the distribution system.

9.1.3.2 Minimize the Amount of Working Time

The days needed for the work in the outage area depends on the number of crews, the workload, and the period needed for each aspect of the work. Also, this depends on the feeder configurations, due to suffering from unnecessary interruptions. The total time needed for the scheduling outage can be expressed as:

$$T_W = \sum_{z=1}^{Z} \frac{W_z}{N_z^{\text{crew}}},$$

where

Z is the number of service zones to be scheduled for outage,

W_z is the total workload of the service zone z with the unit "hours per crew,"

N_z^{crew} is the number of crews in the service zone z.

The planning engineers and managers should balance the crew salaries and time budget to improve work efficiency.

9.1.3.3 Minimize the Span of Time for a Feeder

From the engineering viewpoint, the work for the areas within a distribution feeder should be implemented in the shortest continuous time. The span of time for a feeder in the scheduling outage is formulated as:

$$T_S = \max (T_i^{\text{end}} - T_i^{\text{beg}}), \forall\ i = 1,\ 2,\ 3,\ \cdots,\ N.$$

where

T_i^{end} is the starting time of the zones in feeder i,

T_i^{beg} is the ending time of the zones in feeder i,

N is the number of feeders related to the scheduling outage.

Minimizing the maximum value of the time span for a scheduled outage in a feeder means shortening a single zone's working time.

9.1.3.4 Minimize the Divergence of Working Teams

It is expected that the scheduled outage zones in the same feeder are supervised by a fixed crew team because of practical reasons such as familiarity, locations, and/or customers' recognizing them. The variance of working teams for a feeder is considered by:

$$W_D = \max T_{\text{Fi}}, \forall\ i = 1,\ 2,\ 3,\ \cdots,\ N,$$

where T_{Fi} is the number of teams for the zones in feeder i.

9.2 Switching Procedure for Scheduled Outage

The switching steps are generated based on an automatically generated and heuristic approach. This approach is based on individual experience or other expert systems (rules), which are the enriched library of the database of knowledge related to safety. Before the operators execute the switching steps, steps must be approved by their supervisors on the DMS system. The automatic generation of switching steps is based on the latest topological statuses from the SCADA system. The operators will consider the automatically generated steps as the references because sometimes the topologies of the SCADA database may not be up to date. Verification of switching steps must be carefully reviewed and examined.

The framework of the switching procedure includes:

1. The coordination between the distribution control center and the crew to be dispatched or at the site.

2. Open the closest remote-controlled switches within the boundary to be scheduled.

3. Once these switches are remotely opened by operators, then the crew will determine the smallest area of disconnectors (non-remote-controlled switches) to be open.

4. Once step 3 is executed by the crew, the crew reports to the operators in the control center that it is safe to close those remote-controlled switches.

Four examples with different topologies and complexities will be demonstrated in this section to show the steps of switching procedures for a scheduled outage.

Example 1: A Two-Feeder System

Shown in Fig. 9.2 is a two-feeder example. In this case, e_1 and e_9 are circuit breakers and e_5 is a normally open tie switch. All of these three switches are remote-controlled; e_2 and e_4 are two disconnectors (non-remote-controlled switches). The scheduled outage segment is e_3 and the number of customers serviced by distribution nodes from node 2 to node 9 can be defined as 100, 150, 100, 200, 150, 300, 220, and 200, respectively. As shown in Step 1, the circuit breaker is tripped to isolate this area first. Then, the crew will open e_2 and e_4 to narrow the outage area down to the smallest area. In these two steps, the affected customers number 550 since the nodes 2, 3, 4, and 5 are all isolated. In Step 3, the breaker and the tie switch are closed to restore service for unscheduled areas, and the affected customers will be reduced to 250.

Example 2: A Three-Feeder System

A three-feeder system is shown in Fig. 9.3. e_{10} and e_{16} are two normally open tie switches connected with other feeders or substations. The red (see color e-book) line segments (e_4, e_5, e_7, e_8, e_{11}, and e_{12}) represent disconnectors (non-remote-controlled switches). The scheduled outage segment is e_6 in this case. The row vector to display the number of customers serviced by different distribution nodes is shown as [0 100 130 150 100 200 150 300 170 210 0 200

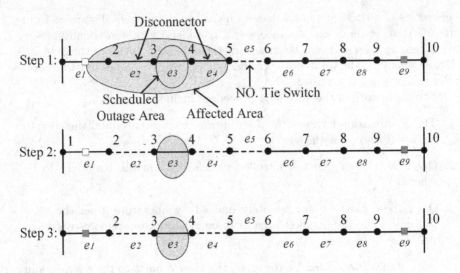

FIGURE 9.2
Switching procedure example for a two-feeder system.

130 220 300 0]. The corresponding position in the row vector represents the nodes' index. In Step 1, the circuit breaker e_1 is tripped (e_{10} and e_{15} stay open) to isolate the whole area first and the affected customers number 2,360 (all customers in this case). Then, the crew will open e_5, e_7, and e_{11} to narrow the outage area down to the smallest area. In Step 3, the breaker e_1 and the tie switches e_{10} and e_{15} are closed to restore service for unscheduled areas. The affected customers will be reduced to 350.

Example 3: A Complex System

As shown in Fig 9.4, Case 3 is a complex but more realistic example. The scheduled outage segment is e_{11}. If we divide this topology into two parts by the scheduled outage segment, two visible tree topologies can be demonstrated. The two normally closed switches (e_3 and e_{19}) are remote-controlled switches. The normally open tie switches (e_{17} and e_{20}) are connected with other feeders or substations. The row vector to display the number of customers serviced by different distribution nodes is shown as [0 100 130 150 100 200 150 300 170 210 110 200 130 220 300 150 120 0 100 0 0]. The corresponding position in the row vector represents the nodes' index. In Step 1, the boundary remote-controlled switches e_3 and e_{19} are opened instead of tripping the breaker to avoid affecting more customers. However, the number of affected customers is still huge (almost all customers are affected). Then, the crew will open e_{10} and e_{12} to narrow the outage area down to the smallest area. In Step 3, the condition preassumes that closing the tie switch e_{20} is a better option to con-

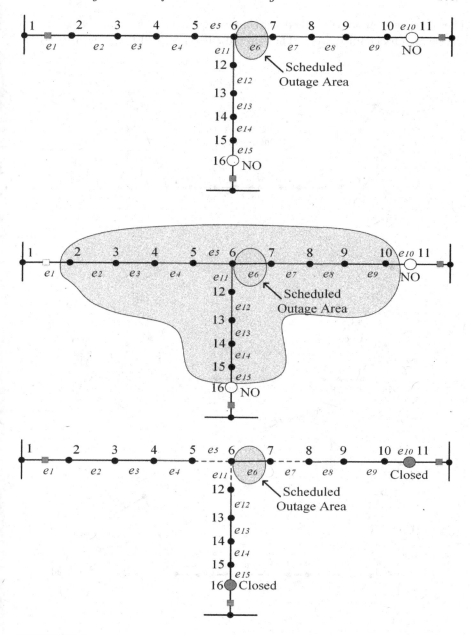

FIGURE 9.3
Switching procedure example for a three-feeder system.

nect adjacent feeders or substations based on the economic and optimization consideration (also, the operator can close e_{17} and e_{19} to restore the service).

Therefore, the remote-controlled switches e_3, e_{19}, and e_{20} are closed to restore service for unscheduled areas. The affected customers will be reduced to 310.

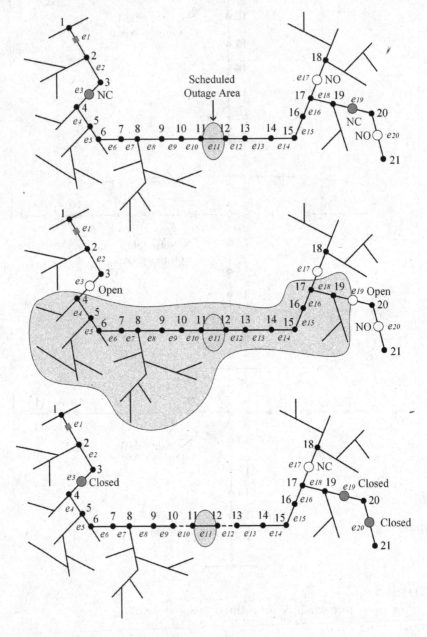

FIGURE 9.4
Switching procedure example for a complex system.

Example 4: A Three-Feeder System with Multiple NO Tie Switches

Fig 9.5 is a three-feeder example and the service areas by different substations have been indicated by different colors (see color e-book). The scheduled outage segment is e_6, which is serviced by feeder 1. The normally open tie switches are indicated as e_{11}, e_{13}, A, B, C, D, and E in this scenario. Two normally closed remote-controlled switches are e_3 and e_9 and the non-remote-controlled switches are e_4, e_5, e_7, e_8, e_{10}, and e_{12}. The row vector to display the number of customers serviced by different distribution nodes is shown as [0 100 130 150 100 200 150 300 170 0 130 0 300 0]. The corresponding position in the row vector represents the nodes' index. In the first step, the boundary remote-controlled switches e_3 and e_9 are opened to isolate the scheduled outage area. The number of affected customers is 1,500. Then, the crew will open e_5 and e_7 in order to reduce the number of affected customers. In the last step, the crew and operators will select a normally open tie switch to close from A, B, C, and D based on a comprehensive consideration (economical and operational) to restore the service in the green area. After service restoration, the affected customers will be reduced to 350.

Detailed information of these four cases with corresponding input variables are shown in Listing 9.1. The MATLAB functions illustrated in Listing 9.2 find out the best switching option. Combined with Listing 8.4, the number of affected customers during the switching procedure can be calculated.

Listing 9.1
Input Variables of 4 Cases for Switching Procedure

```
1   casenumb = input('Input the case number (1-4): ');
2
3   %-------------------------------------------------------------%
4   % Case 1: two feeder example: scheduled switching segment is E3.
5   %-------------------------------------------------------------%
6   if casenumb == 1
7       IncM =[1 0 0 0 0 0 0 0 0
8              1 1 0 0 0 0 0 0 0
9              0 1 1 0 0 0 0 0 0
10             0 0 1 1 0 0 0 0 0
11             0 0 0 1 1 0 0 0 0
12             0 0 0 0 1 1 0 0 0
13             0 0 0 0 0 1 1 0 0
14             0 0 0 0 0 0 1 1 0
15             0 0 0 0 0 0 0 1 1
16             0 0 0 0 0 0 0 0 1];
17
18       % Node 1, 10 are sources (or have power injection from ...
                other feeders).
19       Sourcevec = [1 0 0 0 0 0 0 0 0 1];
20
21       % Initialization of number of customers and connection states.
```

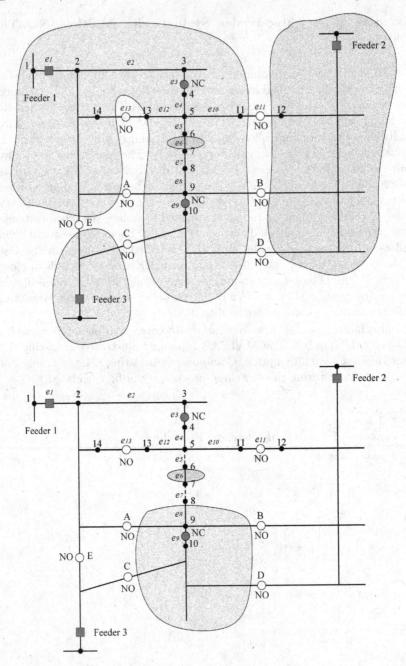

FIGURE 9.5
Switching procedure example for a three-feeder system.

```
22      Customers = [0 100 150 100 200 150 300 220 200 0];
23      onOffVec = [1 1 1 1 1 1 1 1 1];
24
25      % Define switch types: 0-line segment; 1-non-RCS; 2-RCS.
26      sw_type = [2 1 0 1 2 0 0 0 2];
27
28      [Output_step1, Output_step2, onOffVec1, onOffVec2] = ...
            connectstatus(IncM,sw_type,onOffVec);
29
30   %--------------------------------------------------------------%
31   % Case 2: example 1: scheduled switching segment is E6.
32   %--------------------------------------------------------------%
33   elseif casenumb == 2
34      IncM =[1 0 0 0 0 0 0 0 0 0 0 0 0 0 0
35             1 1 0 0 0 0 0 0 0 0 0 0 0 0 0
36             0 1 1 0 0 0 0 0 0 0 0 0 0 0 0
37             0 0 1 1 0 0 0 0 0 0 0 0 0 0 0
38             0 0 0 1 1 0 0 0 0 0 0 0 0 0 0
39             0 0 0 0 1 1 0 0 0 0 1 0 0 0 0
40             0 0 0 0 0 1 1 0 0 0 0 0 0 0 0
41             0 0 0 0 0 0 1 1 0 0 0 0 0 0 0
42             0 0 0 0 0 0 0 1 1 0 0 0 0 0 0
43             0 0 0 0 0 0 0 0 1 1 0 0 0 0 0
44             0 0 0 0 0 0 0 0 0 1 0 0 0 0 0
45             0 0 0 0 0 0 0 0 0 0 1 1 0 0 0
46             0 0 0 0 0 0 0 0 0 0 0 1 1 0 0
47             0 0 0 0 0 0 0 0 0 0 0 0 1 1 0
48             0 0 0 0 0 0 0 0 0 0 0 0 0 1 1
49             0 0 0 0 0 0 0 0 0 0 0 0 0 0 1];
50
51      % Node 1, 11, 16 are sources (or have power injection from ...
            other feeders).
52      Sourcevec = [1 0 0 0 0 0 0 0 0 0 1 0 0 0 0 1];
53
54      % Initialization of number of customers and connection states.
55      Customers = [0 100 130 150 100 200 150 300 170 210 0 200 ...
            130 220 300 0];
56      onOffVec = [1 1 1 1 1 1 1 1 1 1 1 1 1 1 1 1];
57
58      % Define switch types: 0-line segment; 1-non-RCS; 2-RCS.
59      sw_type = [2 0 0 1 1 0 1 1 0 2 1 1 0 0 2];
60
61      [Output_step1, Output_step2, onOffVec1, onOffVec2] = ...
            connectstatus(IncM,sw_type,onOffVec);
62
63   %--------------------------------------------------------------%
64   % Case 3: example 2: scheduled switching segment is E11.
65   %--------------------------------------------------------------%
66   elseif casenumb == 3
67      IncM =[1 0 0 0 0 0 0 0 0 0 0 0 0 0 0 0 0 0 0 0 0
68             1 1 0 0 0 0 0 0 0 0 0 0 0 0 0 0 0 0 0 0 0
69             0 1 1 0 0 0 0 0 0 0 0 0 0 0 0 0 0 0 0 0 0
70             0 0 1 1 0 0 0 0 0 0 0 0 0 0 0 0 0 0 0 0 0
71             0 0 0 1 1 0 0 0 0 0 0 0 0 0 0 0 0 0 0 0 0
72             0 0 0 0 1 1 0 0 0 0 0 0 0 0 0 0 0 0 0 0 0
73             0 0 0 0 0 1 1 0 0 0 0 0 0 0 0 0 0 0 0 0 1
74             0 0 0 0 0 0 1 1 0 0 0 0 0 0 0 0 0 0 0 0 0
```

```
75              0 0 0 0 0 0 0 1 1 0 0 0 0 0 0 0 0 0 0 0
76              0 0 0 0 0 0 0 0 1 1 0 0 0 0 0 0 0 0 0 0
77              0 0 0 0 0 0 0 0 0 1 1 0 0 0 0 0 0 0 0 0
78              0 0 0 0 0 0 0 0 0 0 1 1 0 0 0 0 0 0 0 0
79              0 0 0 0 0 0 0 0 0 0 0 1 1 0 0 0 0 0 0 0
80              0 0 0 0 0 0 0 0 0 0 0 0 1 1 0 0 0 0 0 0
81              0 0 0 0 0 0 0 0 0 0 0 0 0 1 1 0 0 0 0 0
82              0 0 0 0 0 0 0 0 0 0 0 0 0 0 1 1 0 0 0 0
83              0 0 0 0 0 0 0 0 0 0 0 0 0 0 0 1 1 1 0 0
84              0 0 0 0 0 0 0 0 0 0 0 0 0 0 0 0 1 0 0 0
85              0 0 0 0 0 0 0 0 0 0 0 0 0 0 0 0 0 1 1 0
86              0 0 0 0 0 0 0 0 0 0 0 0 0 0 0 0 0 0 1 0
87              0 0 0 0 0 0 0 0 0 0 0 0 0 0 0 0 0 0 0 1];
88
89      % Node 1, 20 are the sources (or have power injection from ...
            other feeders).
90      Sourcevec = [1 0 0 0 0 0 0 0 0 0 0 0 0 0 0 0 0 0 0 1 0];
91
92      % Initialization of number of customers and connection states.
93      Customers = [0 100 130 150 100 200 150 300 170 210 110 200 ...
            130 220 300 150 120 0 100 0 0];
94      onOffVec = [1 1 1 1 1 1 1 1 1 1 1 1 1 1 1 1 1 1 1 1 1];
95
96      % Define switch types: 0-line segment; 1-non-RCS with power ...
             injection;
97      % 2-RCS; 3-RCS without power injection; 4-breaker.
98      sw_type = [4 0 2 0 1 1 1 1 1 1 0 1 1 1 1 1 3 0 2 3];
99
100     % The non-RCS is not close to the target segment.
101     % sw_type = [4 0 2 0 1 1 1 1 1 0 0 1 1 1 1 1 3 0 2 3];
102
103     [Output_step1, Output_step2, onOffVec1, onOffVec2] = ...
            connectstatus(IncM,sw_type,onOffVec);
104
105     %-----------------------------------------------------------%
106     % Case 4: example 6: scheduled switching segment is E6.
107     %-----------------------------------------------------------%
108     elseif casenumb == 4
109       IncM =[1 0 0 0 0 0 0 0 0 0 0 0 0
110             1 1 0 0 0 0 0 0 0 0 0 0 0
111             0 1 1 0 0 0 0 0 0 0 0 0 0
112             0 0 1 1 0 0 0 0 0 0 0 0 0
113             0 0 0 1 1 0 0 0 0 1 0 1 0
114             0 0 0 0 1 1 0 0 0 0 0 0 0
115             0 0 0 0 0 1 1 0 0 0 0 0 0
116             0 0 0 0 0 0 1 1 0 0 0 0 0
117             0 0 0 0 0 0 0 1 1 0 0 0 0
118             0 0 0 0 0 0 0 0 1 0 0 0 0
119             0 0 0 0 0 0 0 0 0 1 1 0 0
120             0 0 0 0 0 0 0 0 0 0 1 0 0
121             0 0 0 0 0 0 0 0 0 0 0 1 1
122             0 0 0 0 0 0 0 0 0 0 0 0 1];
123
124     % Node 1, 10 are sources (or have power injection from ...
            other feeders).
125     Sourcevec = [1 0 0 0 0 0 0 0 0 1 0 0 0 0];
126
```

```
127        % Initialization of number of customers and connection states.
128        Customers = [0 100 130 150 100 200 150 300 170 0 130 0 300 0];
129        onOffVec = [1 1 1 1 1 1 1 1 1 1 1 1 1 1];
130
131        % Define switch types: 0-line segment; 1-non-RCS with power ...
               injection;
132        % 2-RCS; 3-RCS without power injection; 4-breaker.
133        sw_type = [4 0 2 1 1 0 1 1 2 1 3 1 3];
134
135        [Output_step1, Output_step2, onOffVec1, onOffVec2] = ...
               connectstatus(IncM,sw_type,onOffVec);
136  end
137  Affected_customer_state1 = ...
           affcustomer(IncM,onOffVec1,Sourcevec,Customers)
138  Affected_customer_state2 = ...
           affcustomer(IncM,onOffVec2,Sourcevec,Customers)
```

Listing 9.2

Function to Find Out the Best Switching Option

```
 1  function [Output_step1, Output_step2, onOffVec1, onOffVec2] = ...
        connectstatus(IncM, sw_type, onOffVec)
 2
 3  % Find romote controlable and non-controlable switches.
 4  nrcsw = find(sw_type == 1);
 5  rcsw = find(sw_type == 2);
 6
 7  % Step 1: RCS in that region are 'rcsw'.
 8  Output_step1 = rcsw;
 9
10  % Open RCS in that region.
11  for n = 1:length(rcsw)
12      m = rcsw(n);
13      onOffVec(m) = 0;
14  end
15
16  onOffVec1 = onOffVec;
17
18  % Input the index of scheduled switching line segment.
19  swsegment = input('input the index of the scheduled switching ...
        line segment: ');
20
21  % Find the head and tail of the target segment based on ...
        Incidence matrix.
22  seg_head = find(IncM(:,swsegment)==1,1,'first');
23  seg_tail = find(IncM(:,swsegment)==1,1,'last');
24
25  % Find edges connected with head and tail nodes.
26  leftincedge = find(IncM(seg_head,:)==1);
27  rightincedge = find(IncM(seg_tail,:)==1);
28
29  % Index of edges incidence to the target segment.
30  incedge = [leftincedge,rightincedge];
31
32  % Delete the index of target segment.
```

```
33   incedge(incedge == swsegment) = [];
34
35   % Check if the incidence edges are non-controlable switches.
36   open_ncsw = setdiff(incedge,intersect(incedge,nrcsw));
37
38   % The incidence edges are non-controlable switches.
39   if sum(open_ncsw) == 0
40
41       % Step 2: non-RCS in that segment are 'incedge'.
42       Output_step2 = incedge;
43       % Open non-RCS.
44       for i = 1:length(incedge)
45           j = incedge(i);
46           onOffVec(j) = 0;
47       end
48
49   % If not, find the most adjacent non-controlable switch.
50   else
51       for i = 1:length(open_ncsw)
52           j = open_ncsw(i);
53           [value,index] = min(abs(j - nrcsw));
54           onOffVec(nrcsw(index)) = 0;
55       end
56       onOffVec(intersect(incedge,nrcsw)) = 0;
57       % Step 2: non-RCS in that segment are 'incedge'.
58       Output_step2 = [nrcsw(index),intersect(incedge,nrcsw)];
59   end
60
61   % Step 3: Close RCS temproarily.
62   for n = 1:length(rcsw)
63       m = rcsw(n);
64       onOffVec(m) = 1;
65   end
66   onOffVec2 = onOffVec;
```

9.3   Fraud Detection

Energy fraud is a notorious problem in electric power systems, and has serious implications for both utility companies and legitimate users. In the U.S., it is estimated that utility companies lose billions of dollars in revenue every year, while in developing countries, energy theft can amount to 50% of the total energy delivered. Energy fraud also leads to excessive energy consumption, which may cause equipment malfunction or damage and often enables criminal activities, such as the illegal production of substances [185].

Fraud detection can be based on two references: sensors and profiles. Sensors are employed to monitor and supervise meters at the customers' level, the distribution grid, and the communication network. However, the sensor system presents high deployment and maintenance costs, and false alarms may happen during the detection process. Fraud detection based on profile analyzes significant variations in consumption patterns. An approach to combine the profile-based fraud detection method with switching operations of topology reconfiguration can localize the potential fraudulent load.

9.3.1 Profile-Based Fraud Detection

Profile-based fraud detection seeks to detect the irregularity of potentially tampered with data by analyzing abnormal variations in short-period consumption patterns. A common assumption in all related work is the deployment of AMI, or else another real-time meter reading method is performed in the test region. The input dataset to construct the consumption pattern can be obtained from the archived data from the previous short period, which is sufficient to detect ongoing fraud.

9.3.1.1 Threshold of Anomalies

The anomaly threshold of a smart meter identifies whether a specific reading is anomalous. It compares the current consumption curve with the pattern generated from the previous short period. If the difference between two patterns is greater than the user-defined threshold, the reading is anomalous.

For a comparison between the energy supplied by the grid and the energy reported by smart meters, the following equation shows the difference between these two in a subsystem during time interval t,

$$\Delta \varepsilon(t) = \varepsilon_{\text{head}}(t) - \varepsilon_{\text{loss}}(t) - \sum \varepsilon_{\text{user}}(t),$$

where $\varepsilon_{\text{head}}(t)$ is the energy reported by the low-voltage grid meter (DT, FRTU, or the primary reading), $\varepsilon_{\text{loss}}(t)$ is the power losses, and $\varepsilon_{\text{user}}(t)$ is the consumption report of the smart meter from the user side. An abnormal state is triggered when $\Delta \varepsilon(t)$ exceeds the user-defined threshold.

The cost of the fraud detection procedure balanced against the cost of fraudulence can consider another threshold to decide whether or not to execute the following switching procedure to localize the fraud. As shown in the following equation, $p_r(t)$ is the real-time price in the subsystem and $\Delta c(t)$ is the cost of fraudulence

$$\Delta c(t) = \sum p_r(t) \cdot \Delta \varepsilon(t).$$

9.3.1.2 Fraudulence Frequency

The fraudulence frequency of a smart meter is the ratio between the number of anomalous readings and the total number of readings over a test period. Assuming a smart meter obtains the consumption information every τ minutes, the fraudulence frequency f_F is shown as:

$$f_F = \frac{N_a}{\frac{60}{\tau} \cdot 24 \cdot N_d},$$

where N_a is the number of anomalous readings and N_d is the number of days during the test period. For example, assume that a typical smart meter sends out readings every 10 minutes. If utility companies collect one month (30 days) of data and 100 readings during this interval are anomalous, then the fraudulence frequency is around 2.3%. This frequency will be considered at the beginning of the switching procedure to improve searching efficiency.

9.3.2 Switching Strategies

As discussed before, a distribution grid can be modeled by a graph $G = (V, E)$, a set of vertices V, and a set of edges, E. In $G = (V, E)$, substations, distributed energy resources, and distribution loads are referred to as vertices V. The overhead or underground lines with switches are treated as edges E. After converting the topology to an adjacency or incidence matrix, the fraud inference with switching strategies can be performed.

Algorithm 7 is the pseudocode that shows the iterative procedure for how the fraudulent load node is localized for each iteration. The following variables are defined as:

M_i Incidence matrix representing system topology. If vertex i is connected to edge j, then $M_i[i, j] = 1$, otherwise, 0.

M_a Adjacent matrix representing system topology. If vertex i is adjacent to vertex j, then $M_a[i, j] = 1$, otherwise, 0.

V_r Row vector indicating the open/close or fraudulent status of switches/breakers/reclosers with length $|E|$. If switch i is open or has an indicator that shows there is fraud, then $V_r[i] = 0$, otherwise, 1.

V_s Row vector indicating the location of power sources with length $|V|$. If vertex i is a substation-transformer or associated with distribution generation (DG), then $V_s[i] = 1$, otherwise, 0.

V_{frad} Row vector indicating the existence of fraud with length $|V|$. If vertex i has a fraud, then $V_{\text{frad}}[i] = 0$, otherwise, 1.

The algorithm proposed here processes according to the graph-theoretic switching strategies. A preassumption is that a smart meter or fraud indicator is installed in each feeder head. The algorithm starts with balancing the benefits of detecting the fraud and then checking the node with the highest fraudulent frequency in the fraudulent feeder first. The DFS or any other searching method is then performed to decide the following node to be checked. Since the incidence matrix is constructed by vertices and edges, the changing of connection status can be displayed on this matrix directly, and the input

---

Algorithm 7 Fraud Localization Algorithm

Input:
  M_i, V_r, V_s
Iteration Fuction: $F_{\text{frad}}\ (M_i, V_r, V_s)$
 1: $M_i \leftarrow M_i \cdot V_r^{\mathsf{T}}$.
 2: $M_a \leftarrow M_i \cdot M_i^{\mathsf{T}}$.
 3: Replace all diagonal elements of M_a with 0.
 4: % Initialization.
 5: $V_{\text{frad}} \leftarrow V_s$.
 6: $\widehat{V}_{\text{frad}} \leftarrow V_{\text{frad}} \cdot 0$.
 7: % Criteria that the state is the same as the previous iteration.
 8: **while** $V_{\text{frad}} - \widehat{V}_{\text{frad}} \neq 0$ **do**
 9:     $\widehat{V}_{\text{frad}} \leftarrow V_{\text{frad}}$.
10:     $V_{\text{frad}} \leftarrow \widehat{V}_{\text{frad}} \cdot M_a + \widehat{V}_{\text{frad}}$.
11: **end while**
12: Replace all nonzero elements in V_{frad} with 1.
13: **return** V_{frad}.
14: % Localize the fraud.
15: Find all 0 elements in V_{frad}.
16: **while** More than one fraudulent nodes in one feeder, change connection states of switches to generate a new V_{frad}. **do**
17:     Repeat $F_{\text{frad}}\ (M_a, V_r, V_s)$.
18: **end while**

---

topology in Algorithm 7 is the incidence matrix. The MATLAB programming script of the fraud localization is demonstrated in Listing 9.3. The algorithm is summarized as follows:

1. Convert the incidence matrix to the adjacency matrix with current connection status of switches for fraudulent nodes detection;

2. Initialize V_{frad} with V_s and set $\widehat{V}_{\text{frad}}$ as the previous iteration result of V_{frad} and initialize $\widehat{V}_{\text{frad}}$ with 0;

3. Check every node status of each iteration by checking $V_{\text{frad}} = \widehat{V}_{\text{frad}}$;

4. If the status of all nodes is stable, replace all nonzero elements in V_{frad} with 1;

5. For the situation in which there is more than one fraudulent node in one feeder, change the connection states of switches to generate a new V_r and repeat $F_{\text{frad}}\ (M_i, V_r, V_s)$;

6. Return V_{frad};

7. Find 0 element(s) in V_{frad} to localize the fraud.

Listing 9.3
Fraud Localization Function Based on the Incidence Matrix

```
1   % Update "EdgeVec" for switches status.
2   % Update "NodeVec" for FIs' results.
3   function fraud_node_vec = fraudet(incMat, EdgeVec, NodeVec)
4
5   % Convert incidence matrix to adjacency matrix
6   % Update the subsystem's topology.
7
8   IndMat = diag(EdgeVec);
9   incMatr = incMat * IndMat;
10  adjMat = incMatr*incMatr';
11  adjMat = abs(adjMat - diag(diag(adjMat)));
12
13  % Initialize values.
14  fraud_node_vec = NodeVec;
15  fraud_node_old = fraud_node_vec*0;
16
17  % Criteria that if the state is same as the previous iteration.
18  while (length(find(fraud_node_old == ...
        0))-length(find(fraud_node_vec == 0)))≠ 0
19      fraud_node_old = fraud_node_vec;
20
21  % Multiplication based on the topology (Adjacency Matrix).
22      fraud_node_vec = fraud_node_old * adjMat + fraud_node_old;
23      idx = find(fraud_node_vec > 0);
24  % Replace all non-zero elements to 1s.
25  % Find "0" elements.
26      fraud_node_vec(idx) = 1;
27  end
```

Fig. 9.6 is the topology of the distribution test network. There are two substations, four source nodes, and 20 load nodes. The four switches at each feeder head are deployed with a fraud indicator or a smart meter to perform the profile-based fraud detection. The four load nodes with the highest fraudulence frequency at each feeder are indicated by red boxes, and the randomly generated fraud load node is in a green circle (see color e-book). The corresponding spanning tree is shown in Fig. 9.7. The four source nodes are lumped into a single source and the indicator installed in e_4 infers an anomaly that may be subject to fraud.

The input incidence matrix M_i of the topology in Fig. 9.7 with all possible

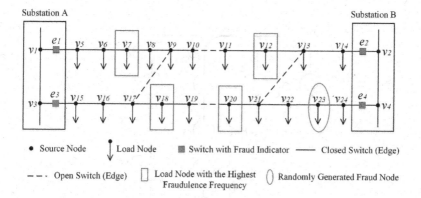

FIGURE 9.6
Topology of distribution test network.

connections is shown as:

	e_1	e_2	e_3	e_4	e_5	e_6	e_7	e_8	e_9	e_{10}	e_{11}	e_{12}	e_{13}	e_{14}	e_{15}	e_{16}	e_{17}	e_{18}	e_{19}	e_{20}	e_{21}	e_{22}	e_{23}	e_{24}
S	1	1	1	1	0	0	0	0	0	0	0	0	0	0	0	0	0	0	0	0	0	0	0	0
v_5	1	0	0	0	1	0	0	0	0	0	0	0	0	0	0	0	0	0	0	0	0	0	0	0
v_6	0	0	0	0	1	1	0	0	0	0	0	0	0	0	0	0	0	0	0	0	0	0	0	0
v_7	0	0	0	0	0	1	1	0	0	0	0	0	0	0	0	0	0	0	0	0	0	0	0	0
v_8	0	0	0	0	0	0	1	1	0	0	0	0	0	0	0	0	0	0	0	0	0	0	0	0
v_9	0	0	0	0	0	0	0	1	1	0	0	0	0	0	0	0	0	0	0	0	1	0	0	0
v_{10}	0	0	0	0	0	0	0	0	1	0	0	0	0	0	0	0	0	0	0	0	0	1	0	0
v_{11}	0	0	0	0	0	0	0	0	0	0	0	1	0	0	0	0	0	0	0	0	1	0	0	0
v_{12}	0	0	0	0	0	0	0	0	0	0	1	1	0	0	0	0	0	0	0	0	0	0	0	0
v_{13}	0	0	0	0	0	0	0	0	0	1	1	0	0	0	0	0	0	0	0	0	0	0	1	0
v_{14}	0	0	0	0	0	0	0	0	0	1	0	0	0	0	0	0	0	0	0	0	0	0	0	0
v_{15}	0	0	0	0	0	0	0	0	0	0	0	0	1	0	0	0	0	0	0	0	0	0	0	0
v_{16}	0	0	0	0	0	0	0	0	0	0	0	0	1	1	0	0	0	0	0	0	0	0	0	0
v_{17}	0	0	0	0	0	0	0	0	0	0	0	0	0	1	1	0	0	0	0	1	0	0	0	0
v_{18}	0	0	0	0	0	0	0	0	0	0	0	0	0	0	1	1	0	0	0	0	0	0	0	0
v_{19}	0	0	0	0	0	0	0	0	0	0	0	0	0	0	0	1	0	0	0	0	0	0	0	1
v_{20}	0	0	0	0	0	0	0	0	0	0	0	0	0	0	0	0	0	0	0	1	0	0	0	1
v_{21}	0	0	0	0	0	0	0	0	0	0	0	0	0	0	0	0	0	1	1	0	0	1	0	0
v_{22}	0	0	0	0	0	0	0	0	0	0	0	0	0	0	0	0	0	1	1	0	0	0	0	0
v_{23}	0	0	0	0	0	0	0	0	0	0	0	0	0	0	0	0	1	1	0	0	0	0	0	0
v_{24}	0	0	0	1	0	0	0	0	0	0	0	0	0	0	0	1	0	0	0	0	0	0	0	0

In all steps, since all source nodes are lumped together as node S and S is the only source node, the power source vector

$$V_s = [1\ 0\ 0\ 0\ 0\ 0\ 0\ 0\ 0\ 0\ 0\ 0\ 0\ 0\ 0\ 0\ 0\ 0\ 0\ 0\ 0\ 0\ 0\ 0].$$

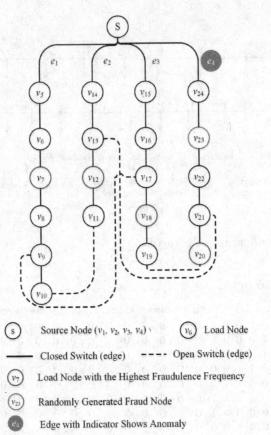

FIGURE 9.7
Converted tree topology of this example.

Step 1: Since v_{20} has the highest fraudulence frequency, the switching procedure starts with checking the state of v_{20}. As shown in Fig. 9.8 (a), we broke the switch between v_{20} and v_{21} (e_{20}) and closed the switch between v_{19} and v_{20} (e_{24}) simultaneously. The value of switch e_4 in V_r is 0. The updated

$$V_r = [1\ 1\ 1\ 0\ 1\ 1\ 1\ 1\ 1\ 1\ 1\ 1\ 1\ 1\ 1\ 1\ 1\ 1\ 1\ 1\ 0\ 0\ 0\ 0\ 1],$$

and the updated M_a can be calculated through

$$M_i = M_i \cdot V_r^\top, \quad M_a = M_i \cdot M_i^\top.$$

Then, the row vector indicating the existence of a fraud

$$V_{\text{frad}} = \widehat{V}_{\text{frad}} \cdot M_a + \widehat{V}_{\text{frad}},$$

and

$$V_{\text{frad}} = [1\ 1\ 1\ 1\ 1\ 1\ 1\ 1\ 1\ 1\ 1\ 1\ 1\ 1\ 1\ 1\ 1\ 1\ 0\ 0\ 0\ 0].$$

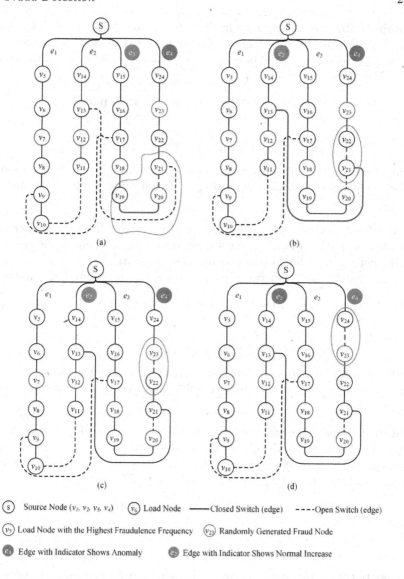

FIGURE 9.8
Graph representation of switching procedures.

Since the current connection status of the topology has been changed, we modified values on corresponding positions in the input incidence matrix. The consumption reading of the indicator installed in e_3 displayed a corresponding increase but within a normal range. The e_4 still showed anomalies, meaning that the fraud was in the load nodes v_{21}, v_{22}, v_{23}, or v_{24}.

Step 2: This step is displayed in Fig. 9.8 (b). According to DFS, we then broke the switch between v_{21} and v_{22} (e_{19}) and closed the switch between v_{13} and v_{21} (e_{23}) as shown in the blue circle (see color e-book) to check the state of v_{21}. The updated

$$V_r = [1\ 1\ 1\ 0\ 1\ 1\ 1\ 1\ 1\ 1\ 1\ 1\ 1\ 1\ 1\ 1\ 1\ 0\ 0\ 0\ 0\ 1\ 1],$$

and

$$V_{\text{frad}} = [1\ 1\ 1\ 1\ 1\ 1\ 1\ 1\ 1\ 1\ 1\ 1\ 1\ 1\ 1\ 1\ 1\ 1\ 0\ 0\ 0].$$

Similiar to Step 1, the consumption reading in e_2 rose due to the additional load in v_{21} while the anomalous reading in e_4 indicated the fraud existed in v_{22}, v_{23}, or v_{24}.

Step 3: As demonstrated in Fig. 9.8 (c), the search follows the connection sequence within the fraudulent feeder, and this step is to check the state of v_{22}. Repeat the switching operation to open the switch between v_{22} and v_{23} (e_{18}) and close the switch between v_{21} and v_{22} (e_{19}). In this step, the updated

$$V_r = [1\ 1\ 1\ 0\ 1\ 1\ 1\ 1\ 1\ 1\ 1\ 1\ 1\ 1\ 1\ 1\ 0\ 1\ 0\ 0\ 0\ 1\ 1],$$

and

$$V_{\text{frad}} = [1\ 1\ 1\ 1\ 1\ 1\ 1\ 1\ 1\ 1\ 1\ 1\ 1\ 1\ 1\ 1\ 1\ 1\ 1\ 0\ 0].$$

The data report obtained from the e_2 feeder added up the load consumption data from v_{22}, but within an acceptable limit. The fraud was detected in either v_{23} or v_{24}.

Step 4: Repeat the switching procedure as in the previous steps. As illustrated in Fig. 9.8 (d), we opened the switch between v_{23} and v_{24} (e_{17}) and closed the switch between v_{22} and v_{23} (e_{18}) in this step. As a result, the fraud indicator installed in e_2 responded with an alarm. The updated

$$V_r = [1\ 0\ 1\ 1\ 1\ 1\ 1\ 1\ 1\ 1\ 1\ 1\ 1\ 1\ 1\ 0\ 1\ 1\ 0\ 0\ 0\ 1\ 1],$$

and

$$V_{\text{frad}} = [1\ 1\ 1\ 1\ 1\ 1\ 1\ 0\ 0\ 0\ 0\ 1\ 1\ 1\ 1\ 1\ 0\ 0\ 0\ 1].$$

Through the cross comparison between V_{frad}s in these four steps, we can decide that the fraud was in the load node v_{23}.

9.4   Conclusions

This chapter explores the scheduled outage for maintenance where it involves recommended safety-related switching steps required for a segment. The switching sequences will be based on the latest topological status and how it can change from one topology state to another with minimal interruption. The sequence of switching also takes into consideration of the safety-related functions, where this ensures complete de-energization prior to closing a switch to energize the other part of the segment from other sources. The fraud inference reconfiguration is also done to determine the deviation for each segment (between the two remote-controlled switches) to observe the change of loading conditions corresponding to each segment. Each step will involve open and closed switching steps simultaneously. Future work will include operational constraints that may involve loop conditions in transition to other feeder(s).

Mini Project 7: Implement the Schemes to Determine Unscheduled and Scheduled Outages

This mini project is an incremental enhancement of the last chapter where it involves iterative search to narrow the search area between boundary switches that are not remote-controllable in the faulted area. Compare the script of Chapter 8 with Chapter 7. See where it fits for the module to involve iteration of bi-section search to determine the steps required to find out the smaller area for a hypothesized fault. Observe how it would affect the reliability indices for the test case.

For scheduled outages (Chapter 9), determine if the selected line segment and the sequence of switching involved to make sense of the automatic generation of boundary switches; in other words that could be between remote-controlled and non-remote-controlled switches. Assess the estimated time and evaluate the SAIDI, SAIFI, and CAIDI indices. Verify the steps for any selected line to be scheduled for maintenance.

Congratulations: if you get this far, you may have successfully implemented the MATLAB version of DMS software!

Bibliography

[1] B. Hannaford. Opportunities and problems in the electric distribution system. *Journal of the A.I.E.E.*, 45(2):180–184, Feb. 1926.

[2] G. M. Burt, J. R. McDonald, A. G. King, J. Spiller, D. Brooke, and R. Samwell. Intelligent on-line decision support for distribution system control and operation. *IEEE Transactions on Power Systems*, 10(4):1820–1827, Nov. 1995.

[3] S. Kazemi, M. Fotuhi-Firuzabad, M. Sanaye-Pasand, and M. Lehtonen. Impacts of automatic control systems of loop restoration scheme on the distribution system reliability. *IET Generation, Transmission Distribution*, 3(10):891–902, Oct. 2009.

[4] Y. Mao and K. Miu. Switch placement to improve system reliability for radial distribution systems with distributed generation. In *IEEE Power Engineering Society General Meeting*, volume 1, page 890, Jun. 2004.

[5] M. I. Abouheaf, W. J. Lee, and F. L. Lewis. Dynamic formulation and approximation methods to solve economic dispatch problems. *IET Generation, Transmission Distribution*, 7(8):866–873, Aug. 2013.

[6] S. M. Dean. The design and operation of a metropolitan electrical system from the viewpoint of possible major shutdowns. *Electrical Engineering*, 59(10):575–579, Oct. 1940.

[7] J. P. Britton. An open, object-based model as the basis of an architecture for distribution control centers. *IEEE Transactions on Power Systems*, 7(4):1500–1508, Nov. 1992.

[8] H.-J. Lee and Y.-M. Park. A restoration aid expert system for distribution substations. *IEEE Transactions on Power Delivery*, 11(4):1765–1770, Oct. 1996.

[9] E. A. Sims. Distribution system operation and maintenance. *Electronics and Power*, 12(2):51–54, Feb. 1966.

[10] B. W. McConnell, T. W. Reddoch, S. L. Purucker, and L. D. Monteen. Distribution energy control center experiment. *IEEE Power Engineering Review*, PER-3(6):33, Jun. 1983.

[11] A. K. Subramanian and J. Wilbur. Power system security functions of the energy control center at the Orange and Rockland utilities. *IEEE Power Engineering Review*, PER-3(12):36–37, Dec. 1983.

[12] N. Singh, E. Kliokys, H. Feldmann, R. Kussel, R. Chrustowski, and C. Joborowicz. Power system modelling and analysis in a mixed energy management and distribution management system. *IEEE Transactions on Power Systems*, 13(3):1143–1149, Aug. 1998.

[13] J. S. Lawler, J. S. Lai, L. D. Monteen, J. B. Patton, and D. T. Rizy. Impact of automation on the reliability of the Athens utilities board's distribution system. *IEEE Transactions on Power Delivery*, 4(1):770–778, Jan. 1989.

[14] T. C. Matty. Advanced energy management for home use. *IEEE Transactions on Consumer Electronics*, 35(3):584–588, Aug. 1989.

[15] J. H. Yoon, R. Baldick, and A. Novoselac. Dynamic demand response controller based on real-time retail price for residential buildings. *IEEE Transactions on Smart Grid*, 5(1):121–129, Jan. 2014.

[16] G. C. Heffner and P. Ahlstrand. Systems integration of real-time pricing at PG and E. *IEEE Transactions on Power Systems*, 2(4):1104–1109, Nov. 1987.

[17] R. Mabry and D. Biagini. Telemetering system supports load curtailment and billing. *IEEE Computer Applications in Power*, 6(2):40–45, Apr. 1993.

[18] K. N. Miu, H. D. Chiang, B. Yuan, and G. Darling. Fast service restoration for large-scale distribution systems with priority customers and constraints. *IEEE Transactions on Power Systems*, 13(3):789–795, Aug. 1998.

[19] R. A. Abdoo, G. Lokken, and R. F. Bischke. Load management implementation: Decisions, opportunities and operation. *IEEE Power Engineering Review*, PER-2(10):45–45, Oct. 1982.

[20] S. Buchanan, R. Taylor, S. Paulos, W. Warren, and J. Hay. The electricity consumption impacts of commercial energy management systems. *IEEE Transactions on Power Systems*, 4(1):213–219, Feb. 1989.

[21] M. R. Irving and M. J. H. Sterling. Substation data validation. *IEE Proceedings C - Generation, Transmission and Distribution*, 129(3):119–122, May 1982.

[22] D. J. Gaushell and H. T. Darlington. Supervisory control and data acquisition. *Proceedings of the IEEE*, 75(12):1645–1658, Dec. 1987.

[23] D. J. Gaushell, W. L. Frisbie, and M. H. Kuchefski. Analysis of analog data dynamics for supervisory control and data acquisition systems. *IEEE Power Engineering Review*, PER-3(2):20–21, Feb. 1983.

[24] S.-J. Huang and C.-C. Lin. Application of atm-based network for an integrated distribution SCADA-GIS system. *IEEE Transactions on Power Systems*, 17(1):80–86, Feb. 2002.

[25] K. Ghoshal. Distribution automation: SCADA integration is key. *IEEE Computer Applications in Power*, 10(1):31–35, Jan. 1997.

[26] A. Bruce and R. Lee. A framework for the specification of SCADA data links. *IEEE Transactions on Power Systems*, 9(1):560–564, Feb. 1994.

[27] D. J. Gaushell and W. R. Block. SCADA communication techniques and standards. *IEEE Computer Applications in Power*, 6(3):45–50, Jul. 1993.

[28] A. P. Alves da Silva, V. H. Quintana, and G. K. H. Pang. Solving data acquisition and processing problems in power systems using a pattern analysis approach. *IEE Proceedings C - Generation, Transmission and Distribution*, 138(4):365–376, Jul. 1991.

[29] H. M. Khodr, J. Molea, I. Garcia, C. Hidalgo, P. C. Paiva, J. M. Yusta, and A. J. Urdaneta. Standard levels of energy losses in primary distribution circuits for SCADA application. *IEEE Transactions on Power Systems*, 17(3):615–620, Aug. 2002.

[30] E. K. Chan and H. Ebenhoh. The implementation and evolution of a SCADA system for a large distribution network. *IEEE Transactions on Power Systems*, 7(1):320–326, Feb. 1992.

[31] S. C. Sciacca and W. R. Block. Advanced SCADA concepts. *IEEE Computer Applications in Power*, 8(1):23–28, Jan. 1995.

[32] U. R. Siddiqi and D. L. Lubkeman. An automated strategy for the processing and analysis of distribution automation data. *IEEE Transactions on Power Delivery*, 6(3):1216–1223, Jul. 1991.

[33] M. L. Chan and W. H. Crouch. An integrated load management, distribution automation and distribution SCADA system for Old Dominion Electric Cooperative. *IEEE Transactions on Power Delivery*, 5(1):384–390, Jan. 1990.

[34] Y. S. Sherif and S. S. Zahir. Communication systems for load management. *IEEE Transactions on Power Apparatus and Systems*, PAS-104(12):3329–3337, Dec. 1985.

[35] A. G. Bruce. Reliability analysis of electric utility SCADA systems. *IEEE Transactions on Power Systems*, 13(3):844–849, Aug. 1998.

[36] T. E. Dy-Liacco. Modern control centers and computer networking. *IEEE Computer Applications in Power*, 7(4):17–22, Oct. 1994.

[37] J. Marcuse, B. Menz, and J. R. Payne. Servers in SCADA applications. *IEEE Transactions on Industry Applications*, 33(5):1295–1299, Sept. 1997.

[38] J. T. Powers, R. W. Anderson, and B. A. Smith. Flexible billing tools meet customer demands. *IEEE Computer Applications in Power*, 11(1):54–58, Jan. 1998.

[39] G. D. Friedlander. Power: Matching utility output to customer demand: load management includes the design of energy-supply facilities to meet customer load, plus the control of a utility's "load shape". *IEEE Spectrum*, 13(9):51–53, Sept. 1976.

[40] S. Finster and I. Baumgart. Privacy-aware smart metering: A survey. *IEEE Communications Surveys Tutorials*, 17(2):1088–1101, Second quarter 2015.

[41] W. H. Kersting. *Distribution System Modeling and Analysis*. CRC Press, Taylor & Francis Group, LLC. Boca Raton, FL, USA, 2011.

[42] K. Mehlhorn and P. Sanders. *Algorithms and Data Structures: The Basic Toolbox*. Springer-Verlag, Heidelberg, 2008.

[43] C. Gary and O. R. Oellermann. *Applied and Algorithmic Graph Theory*. McGraw-Hill, New York, USA, 1993.

[44] S. Even. *Graph Algorithms*. Computer Science Press, Maryland, USA, 1979.

[45] A. Gibbons. *Algorithmic Graph Theory*. Cambridge University Press, Cambridge, 1985.

[46] K. Kanoun, D. Atienza, N. Mastronarde, and M. van der Schaar. A unified online directed acyclic graph flow manager for multicore schedulers. In *19th Asia and South Pacific Design Automation Conf. (ASP-DAC)*, pages 714–719, Jan. 2014.

[47] P. Tsiakas, A. Dor, K. Voudouris, and M. Hadjinicolaou. Load balancing in limited intra-cell interference (LICI) networks based on maximum graph-flow algorithms. In *Intl. Conf. on Ultra Modern Telecommunications Workshops*, pages 1–5, Oct. 2009.

[48] J. Li, X.-Y. Ma, C.-C. Liu, and K. P. Schneider. Distribution system restoration with microgrids using spanning tree search. *IEEE Transactions on Power Systems*, 29(6):3021–3029, Nov. 2014.

[49] C. C. Douglas and B. F. Smith. Using symmetries and antisymmetries to analyze a parallel multigrid algorithm: the elliptic boundary value case. *SIAM Journal on Numerical Analysis*, 26(6):1439–1461, Dec. 1989.

[50] S. M. Chan and V. Brandwajn. Partial matrix refactorization. *IEEE Transactions on Power Systems*, 1(1):193–199, Feb. 1986.

[51] S. Pissanetzky. *Sparse Matrix Technology*. Academic Press, Cambridge, MA, 1984.

[52] J. R. Gilbert, C. Moler, and R. Schreiber. Sparse matrices in Matlab: Design and implementation. *SIAM Journal on Matrix Analysis and Applications*, 13(1):333–356, Jul. 1992.

[53] A. E. Brouwer, Arjeh M. Cohen, and A. Neumaier. *Distance-Regular Graphs*. Springer-Verlag, New York, USA, 1989.

[54] K. Das. The Laplacian spectrum of graphs. *Computers & Mathematics with Applications*, 48(5-6):715–724, May 2004.

[55] R. Merris. Laplacian matrices of graphs: A survey. *Linear Algebra and its Applications*, 197–198:143–176, Jan. 2004.

[56] F. Drfler and F. Bullo. Kron reduction of graphs With applications to electrical networks. *IEEE Transactions on Circuits and Systems*, 60(1):150–163, Feb. 2013.

[57] M. T. Goodrich, R. Tamassia, and M. H. Goldwasser. *Data Structures and Algorithms in Java*. John Wiley & Sons, Inc. New York, USA, 2014.

[58] A. B. Kahn. Topological sorting of large networks. *Communications of the ACM*, 5(11):558–562, Nov. 1962.

[59] B. Brown. Section 5: System Arrangements. http://static.schneider-electric.us/assets/consultingengineer/appguidedocs/section5_0307.pdf., 2017.

[60] W. Sadiq and M. E. Orlowska. Analyzing process models using graph reduction techniques. *Information Systems*, 25(2):117–134, Apr. 2000.

[61] H. Lin, Z. Zhao, H. Li, and Z. Chen. A novel graph reduction algorithm to identify structural conflicts. In *35th Annual Hawaii Intl. Conf. on System Sciences*, pages 1–10, Jan. 2002.

[62] W. Sadiq and M. E. Orlowska. Applying graph reduction techniques for identifying structural conflicts in process models. In *11th Intl. Conf. on CAiSE*, pages 195–209, Jun. 1999.

[63] F. S. Yao, X. Q. Zhang, Y. Zhang, and T. H. Wang. Computer decision-making support system for power distribution network planning based on geographical information system. In *China Intl. Conf. on Electricity Distribution*, pages 1–6, Dec. 2008.

[64] O. Huisman and R. A. de By. *Principles of Geographic Information System*. The Intl. Institute for Geo-Information Science and Earth Observation (ITC), Jan. 2001.

[65] K. Zhang, S. Zhang, B. Huang, and X. Ma. Research on integration technology between distribution automation system and geographical information system. In *Asia-Pacific Power and Energy Engineering Conf.*, pages 1–4, Mar. 2009.

[66] D. Leiva, C. Araya, G. Valverde, and J. Quirs-Torts. Statistical representation of demand for gis-based load profile allocation in distribution networks. In *IEEE Manchester PowerTech*, pages 1–6, Jun. 2017.

[67] T. Mohar, K. Bakie, and J. Curk. Advanced planning procedure and operation of distribution network supported by SCADA and GIS. In *IEEE Power Engineering Society Winter Meeting. Conf. Proceedings (Cat. No.00CH37077)*, volume 4, pages 2780–2785, Jan. 2000.

[68] P. Bhargava, A. Jain, S. Singh, and V. K. Thakur. GIS-SCADA: Integration and applications in distribution network. In *Intl. Conf. on Computation of Power, Energy, Information and Communication (IC-CPEIC)*, pages 0354–0357, Apr. 2015.

[69] L. Qi, C. Wang, W. Zhou, and Z. Yang. Design of distribution SCADA system based on open source gis. In *4th Intl. Conf. on Electric Utility Deregulation and Restructuring and Power Technologies (DRPT)*, pages 523–526, Jul. 2011.

[70] MaCGD and NRE. GIS vs Geospatial. https://www.mygeoportal.gov.my/gis-vs-geospatial., 2018.

[71] National Geographic Society. GIS (geographic information system). https://www.nationalgeographic.org/encyclopedia/geographic-information-system-gis/., 2018.

[72] J. E. Campbell and M. Shin. *Geographic Information System Basics*. Academic Press, Cambridge, MA, Dec. 2009.

[73] ESRI. What are tables and attribute information? http://desktop.arcgis.com/en/arcmap/latest/manage-data/tables/what-are-tables-and-attribute-information.htm., 2018.

[74] M. P. Szudzik. The Rosenberg-Strong pairing function. *CoRR*, abs/1706.04129, Jul. 2017.

[75] B. Reddaiah. Enciphering using elegant pairing functions and logical operators. *Intl. Journal of Computer Applications*, 150(9):7–12, Sept. 2016.

[76] F. Stephan. *Recursion Theory*. Departments of Mathematics and Computer Science, National University of Singapore, Oct. 2012.

[77] K. Ruohonen. *Graph Theory*. Swamy & Thulasiraman, Oklahoma, USA, 2013.

[78] S. Bruno, S. Lamonaca, G. Rotondo, U. Stecchi, and M. La Scala. Unbalanced three-phase optimal power flow for smart grids. *IEEE Transactions on Industrial Electronics*, 58(10):4504–4513, Oct. 2011.

[79] T. H. Chen, M. S. Chen, K. J. Hwang, P. Kotas, and E. A. Chebli. Distribution system power flow analysis-a rigid approach. *IEEE Transactions on Power Delivery*, 6(3):1146–1152, Jul. 1991.

[80] M. F. AlHajri and M. E. El-Hawary. Exploiting the radial distribution structure in developing a fast and flexible radial power flow for unbalanced three-phase networks. *IEEE Transactions on Power Delivery*, 25(1):378–389, Jan. 2010.

[81] R. Stoicescu, K. Miu, C. O. Nwankpa, D. Niebur, and Xiaoguang Yang. Three-phase converter models for unbalanced radial power flow studies. *IEEE Transactions on Power Systems*, 17(4):1016–1021, Nov. 2002.

[82] M. Abdel-Akher, M. E. Ahmad, R. N. Mahanty, and K. M. Nor. An approach to determine a pair of power flow solutions related to the voltage stability of unbalanced three-phase networks. *IEEE Transactions on Power Systems*, 23(3):1249–1257, Aug. 2008.

[83] M. E. Galey. Benefits of performing unbalanced voltage calculations. *IEEE Transactions on Industry Applications*, 24(1):15–24, Jan. 1988.

[84] D. Reichelt, E. Ecknauer, and H. Glavitsch. Estimation of steady-state unbalanced system conditions combining conventional power flow and fault analysis software. *IEEE Transactions on Power Systems*, 11(1):422–427, Feb. 1996.

[85] J. C. M. Vieira, W. Freitas, and A. Morelato. Phase-decoupled method for three-phase power flow analysis of unbalanced distribution systems. *IEE Proceedings - Generation, Transmission and Distribution*, 151(5):568–574, Sept. 2004.

[86] G. A. E.-A. Mahmoud and E. S. S. Oda. Investigation of connecting wind turbine to radial distribution system on voltage stability using si index and λ-v curves. *Smart Grid and Renewable Energy*, 7(1):16–45, Jan. 2016.

[87] C. W. Brice. Voltage-drop calculations and power flow studies for rural electric distribution lines. In *34th Annual Conf. on Rural Electric Power*, pages B2/1–B2/8, Apr. 1990.

[88] G. Carpinelli, D. Lauria, and P. Varilone. Voltage stability analysis in unbalanced power systems by optimal power flow. *IEE Proceedings - Generation, Transmission and Distribution*, 153(3):261–268, May 2006.

[89] W.-M. Lin, Y.-S. Su, H.-C. Chin, and J.-H. Teng. Three-phase unbalanced distribution power flow solutions with minimum data preparation. *IEEE Transactions on Power Systems*, 14(3):1178–1183, Aug. 1999.

[90] M. Z. Kamh and R. Iravani. Unbalanced model and power flow analysis of microgrids and active distribution systems. *IEEE Transactions on Power Delivery*, 25(4):2851–2858, Oct. 2010.

[91] Y. Guo, C. Ten, and P. Jirutitijaroen. Online data validation for distribution operations against cybertampering. *IEEE Transactions on Power Systems*, 29(2):550–560, Mar. 2014.

[92] J. S. Wu, K. L. Tomsovic, and C. S. Chen. A heuristic search approach to feeder switching operations for overload, faults, unbalanced flow and maintenance. *IEEE Transactions on Power Delivery*, 6(4):1579–1586, Oct. 1991.

[93] H. D. Chiang, J. C. Wang, and K. N. Miu. Explicit loss formula, voltage formula and current flow formula for large-scale unbalanced distribution systems. *IEEE Transactions on Power Systems*, 12(3):1061–1067, Aug. 1997.

[94] B. Quartey, D. Shaw, and P. Waked. An application of PLC's as an RTU in SCADA systems. In *39th Annual Petroleum and Chemical Industry Conf.*, pages 271–274, Sept. 1992.

[95] P. Kumar, V. K. Chandna, and M. S. Thomas. Intelligent algorithm for preprocessing multiple data at RTU. *IEEE Transactions on Power Systems*, 18(4):1566–1572, Nov. 2003.

[96] J. Ge, L. Tong, Q. Chen, G. Han, and Z. Tang. Unmanned substations employ multimedia network RTUs. *IEEE Computer Applications in Power*, 15(2):36–40, Apr. 2002.

[97] H. L. Smith and T. J. Modzelewski. Enhancing energy management systems with advanced RTU capabilities. *IEEE Computer Applications in Power*, 2(4):26–29, Oct. 1989.

[98] H. L. Smith and W. R. Block. RTUs slave for supervisory systems (power systems). *IEEE Computer Applications in Power*, 6(1):27–32, Jan. 1993.

[99] A. Sanchez, D. Li, Zeyar Aung, and J. R. Williams. Cybersecurity of smart meters. *Smart Grid Handbook*, pages 1–20, Feb. 2016.

[100] D. Moneta and G. Mauri. Factors influencing adoption of smart meters. *Smart Grid Handbook*, pages 1–21, Aug. 2016.

[101] T. Wolf. Meter data collection, management, and analysis. *Smart Grid Handbook*, pages 1–26, Aug. 2016.

[102] S. J. Darby. Smart meters and residential customers. *Smart Grid Handbook*, pages 1–13, Aug. 2016.

[103] Honeywell. PowerSpring Meter Data Management System. https://www.honeywellprocess.com/library/marketing/tech-specs/PIN_PowerSpring.pdf., 2013.

[104] B. Venkatesh and R. Ranjan. Data structure for radial distribution system load flow analysis. *IEE Proceedings - Generation, Transmission and Distribution*, 150(1):101–106, Jan. 2003.

[105] R. Parasher. Load flow analysis of radial distribution network using linear data structure. *CoRR*, abs/1403.4702, 2014.

[106] A. Dukpa, B. Venkatesh, and L. Chang. Radial distribution system analysis using data structure in Matlab environment. In *Large Engineering Systems Conf. on Power Engineering*, pages 284–289, Oct. 2007.

[107] R. Berg, E. S. Hawkins, and W. W. Pleines. Mechanized calculation of unbalanced load flow on radial distribution circuits. *IEEE Transactions on Power Apparatus and Systems*, PAS-86(4):415–421, Apr. 1967.

[108] Y.-Y. Hong and F.-M. Wang. Investigation of impacts of different three-phase transformer connections and load models on unbalance in power systems by optimization. *IEEE Transactions on Power Systems*, 12(2):689–697, May 1997.

[109] W.-M. Lin and H.-C. Chin. A new approach for distribution feeder reconfiguration for loss reduction and service restoration. *IEEE Transactions on Power Delivery*, 13(3):870–875, Jul. 1998.

[110] P. Ravibabu, M. V. S. Ramya, R. Sandeep, M. V. Karthik, and S. Harsha. Implementation of improved genetic algorithm in distribution system with feeder reconfiguration to minimize real power losses. In *2nd Intl. Conf. on Computer Engineering and Technology*, volume 4, pages V4-320–V4-323, Apr. 2010.

[111] M. A. Kashem, V. Ganapathy, and G. B. Jasmon. Network reconfiguration for load balancing in distribution networks. *IEE Proceedings - Generation, Transmission and Distribution*, 146(6):563–567, Nov. 1999.

[112] M. A. Kashem, V. Ganapathy, and G. B. Jasmon. Network reconfiguration for enhancement of voltage stability in distribution networks. *IEE Proceedings - Generation, Transmission and Distribution*, 147(3):171–175, May 2000.

[113] T. P. Wagner, A. Y. Chikhani, and R. Hackam. Feeder reconfiguration for loss reduction: an application of distribution automation. *IEEE Transactions on Power Delivery*, 6(4):1922–1933, Oct. 1991.

[114] S. Civanlar, J. J. Grainger, H. Yin, and S. S. H. Lee. Distribution feeder reconfiguration for loss reduction. *IEEE Transactions on Power Delivery*, 3(3):1217–1223, Jul. 1988.

[115] Y. Y. Hsu and J. H. Yi. Planning of distribution feeder reconfiguration with protective device coordination. *IEEE Transactions on Power Delivery*, 8(3):1340–1347, Jul. 1993.

[116] S. K. Goswami and S. K. Basu. A new algorithm for the reconfiguration of distribution feeders for loss minimization. *IEEE Transactions on Power Delivery*, 7(3):1484–1491, Jul. 1992.

[117] Q. Zhou, D. Shirmohammadi, and W. H. E. Liu. Distribution feeder reconfiguration for service restoration and load balancing. *IEEE Transactions on Power Systems*, 12(2):724–729, May 1997.

[118] T. Taylor and D. Lubkeman. Implementation of heuristic search strategies for distribution feeder reconfiguration. *IEEE Transactions on Power Delivery*, 5(1):239–246, Jan. 1990.

[119] I. Roytelman, V. Melnik, S. S. H. Lee, and R. L. Lugtu. Multi-objective feeder reconfiguration by distribution management system. *IEEE Transactions on Power Systems*, 11(2):661–667, May 1996.

[120] M. E. Baran and F. F. Wu. Network reconfiguration in distribution systems for loss reduction and load balancing. *IEEE Transactions on Power Delivery*, 4(2):1401–1407, Apr. 1989.

[121] J. M. Undrill and T. E. Kostyniak. Advanced power system fault analysis method. *IEEE Transactions on Power Apparatus and Systems*, 94(6):2141–2150, Nov. 1975.

[122] Y. Zhu, Y. H. Yang, B. W. Hogg, W. Q. Zhang, and S. Gao. An expert system for power systems fault analysis. *IEEE Transactions on Power Systems*, 9(1):503–509, Feb. 1994.

[123] J. R. McDonald, G. M. Burt, and D. J. Young. Alarm processing and fault diagnosis using knowledge based systems for transmission and distribution network control. *IEEE Transactions on Power Systems*, 7(3):1292–1298, Aug. 1992.

[124] R. Apel, C. Jaborowicz, and R. Kussel. Fault management in electrical distribution networks. In *16th Intl. Conf. and Exhibition on Electricity Distribution, Part 1: Contributions. CIRED.*, volume 3, pages 1–5, Jun. 2001.

[125] H. E. Brown, C. E. Person, L. K. Kirchmayer, and G. W. Stagg. Digital calculation of 3-phase short circuits by matrix method. *Transactions of the American Institute of Electrical Engineers. Part III: Power Apparatus and Systems*, 79(3):1277–1281, Apr. 1960.

[126] A. H. El-Abiad. Digital calculation of line-to-ground short circuits by matrix method. *Transactions of the American Institute of Electrical Engineers. Part III: Power Apparatus and Systems*, 79(3):323–331, Apr. 1960.

[127] P. M. Anderson. Analysis of simulatenous faults by two-port network theory. *IEEE Transactions on Power Apparatus and Systems*, PAS-90(5):2199–2205, Sept. 1971.

[128] P. M. Anderson, D. W. Bowen, and A. P. Shah. An indefinite admittance network description for fault computation. *IEEE Transactions on Power Apparatus and Systems*, PAS-89(6):1215–1219, Jul. 1970.

[129] A. Balouktsis, D. Tsanakas, and G. Vachtsevanos. Probabilistic short circuit analysis by Monte Carlo simulations and analytical methods. *IEEE Transactions on Power Systems*, 1(3):135–140, Aug. 1986.

[130] X. Zhang, F. Soudi, D. Shirmohammadi, and C. S. Cheng. A distribution short circuit analysis approach using hybrid compensation method. *IEEE Transactions on Power Systems*, 10(4):2053–2059, Nov. 1995.

[131] W. H. Kersting and W. H. Phillips. Distribution system short circuit analysis. In *Proceedings of the 25th Intersociety Energy Conversion Engineering Conf.*, volume 1, pages 310–315, Aug. 1990.

[132] W. H. Kersting and G. Shirek. Short circuit analysis of IEEE test feeders. In *PES T& D*, pages 1–9, May 2012.

[133] T. H. Chen, M. S. Chen, W. J. Lee, P. Kotas, and P. Van Olinda. Distribution system short circuit analysis-a rigid approach. In *Power Industry Computer Application Conf.*, pages 22–28, May 1991.

[134] R. R. P. Sinha. Evaluation of short circuit strength of distribution transformers. *IEEE Transactions on Power Apparatus and Systems*, PAS-101(7):2249–2259, Jul. 1982.

[135] A. Tan, W. H. E. Liu, and D. Shirmohammadi. Transformer and load modeling in short circuit analysis for distribution systems. *IEEE Transactions on Power Systems*, 12(3):1315–1322, Aug. 1997.

[136] F. J. Muench and G. A. Wright. Fault indicators: Types, strengths applications. *IEEE Transactions on Power Apparatus and Systems*, PAS-103(12):3688–3693, Dec. 1984.

[137] H. Thomas and J. Kalmuck. Reset characteristics of automatic resetting type fault indicators on three-phase urd systems. *IEEE Transactions on Power Apparatus and Systems*, PAS-101(8):2439–2442, Aug. 1982.

[138] G. E. Henry. A method for economic evaluation of field failures such as low-voltage side lightning surge failure of distribution transformers. *IEEE Transactions on Power Delivery*, 3(2):813–818, Apr. 1988.

[139] J. H. Teng, W. H. Huang, and S. W. Luan. Automatic and fast faulted line-section location method for distribution systems based on fault indicators. *IEEE Transactions on Power Systems*, 29(4):1653–1662, Jul. 2014.

[140] C. Y. Ho, T. E. Lee, and C. H. Lin. Optimal placement of fault indicators using the immune algorithm. *IEEE Transactions on Power Systems*, 26(1):38–45, Feb. 2011.

[141] W. F. Usida, D. V. Coury, R. A. Flauzino, and I. N. da Silva. Efficient placement of fault indicators in an actual distribution system using evolutionary computing. *IEEE Transactions on Power Systems*, 27(4):1841–1849, Nov. 2012.

[142] A. Shahsavari, S. M. Mazhari, A. Fereidunian, and H. Lesani. Fault indicator deployment in distribution systems considering available control and protection devices: A multi-objective formulation approach. *IEEE Transactions on Power Systems*, 29(5):2359–2369, Sept. 2014.

[143] G. E. Hager, R. N. Medicine Bear, and A. S. Baum. Automated distribution fault locating system. *IEEE Transactions on Industry Applications*, 32(3):704–708, May 1996.

[144] A. L. Morelato and A. J. Monticelli. Heuristic search approach to distribution system restoration. *IEEE Transactions on Power Delivery*, 4(4):2235–2241, Oct. 1989.

[145] C. C. Liu, S. J. Lee, and S. S. Venkata. An expert system operational aid for restoration and loss reduction of distribution systems. *IEEE Transactions on Power Systems*, 3(2):619–626, May 1988.

[146] H. F. Habib, T. Youssef, M. H. Cintuglu, and O. A. Mohammed. Multi-agent-based technique for fault location, isolation, and service restoration. *IEEE Transactions on Industry Applications*, 53(3):1841–1851, May 2017.

[147] Y. S. Ko, T. K. Kang, H. Y. Park, H. Y. Kim, and H. S. Nam. The FRTU-based fault-zone isolation method in the distribution systems. *IEEE Transactions on Power Delivery*, 25(2):1001–1009, Apr. 2010.

[148] P. M. Sonwane, D. P. Kadam, and B. E. Kushare. Distribution system reliability through reconfiguration, fault location, isolation and restoration. In *Intl. Conf. on Control, Automation, Communication and Energy Conservation*, pages 1–6, Jun. 2009.

[149] J. Konsti. Developing fault management in a Distribution Management System based on requirements of Finnish Distribution System Operators. `https://dspace.cc.tut.fi/dpub/bitstream/handle/123456789/25230/Konsti.pdf?isAllowed=y&sequence=1.`, Mar. 2017.

[150] N. Perera, A. D. Rajapakse, and T. E. Buchholzer. Isolation of faults in distribution networks with distributed generators. *IEEE Transactions on Power Delivery*, 23(4):2347–2355, Oct. 2008.

[151] N. D. R. Sarma, V. C. Prasad, K. S. Prakasa Rao, and V. Sankar. A new network reconfiguration technique for service restoration in distribution networks. *IEEE Transactions on Power Delivery*, 9(4):1936–1942, Oct. 1994.

[152] R. E. Brown and A. P. Hanson. Impact of two-stage service restoration on distribution reliability. *IEEE Transactions on Power Systems*, 16(4):624–629, Nov. 2001.

[153] Y. Y. Hsu, H. M. Huang, H. C. Kuo, S. K. Peng, C. W. Chang, K. J. Chang, H. S. Yu, C. E. Chow, and R. T. Kuo. Distribution system service restoration using a heuristic search approach. *IEEE Transactions on Power Delivery*, 7(2):734–740, Apr. 1992.

[154] K. N. Miu, H.-D. Chiang, and R. J. McNulty. Multi-tier service restoration through network reconfiguration and capacitor control for large-scale radial distribution networks. *IEEE Transactions on Power Systems*, 15(3):1001–1007, Aug. 2000.

[155] N. D. R. Sarma, S. Ghosh, K. S. Prakasa Rao, and M. Srinivas. Real time service restoration in distribution networks-a practical approach. *IEEE Transactions on Power Delivery*, 9(4):2064–2070, Oct. 1994.

[156] K. Kawahara, H. Sasaki, J. Kubokawa, H. Asahara, and K. Sugiyama. A proposal of a supporting expert system for outage planning of electric power facilities retaining high power supply reliability. *IEEE Transactions on Power Systems*, 13(4):1453–1458, Nov. 1998.

[157] Z. Xu and X. Li. The construction of interconnected communication system among smart grid and a variety of networks. In *Asia-Pacific Power and Energy Engineering Conf.*, pages 1–5, Mar. 2010.

[158] O. Pauzet. The future of smart metering: The case for public cellular communications. *Metering Intl. Issue*, 3:38–39, Mar. 2011.

[159] G. Deng, S. Fu, K.-I. Shu, and J. Chen. Discussion on advanced metering infrastructure. In *Proc. Electrical Measurement and Instrumentation*, volume 47, Wuhan, China, Jun. 2010.

[160] M. Music, A Bosovic, N. Hasanspahic, S. Avdakovic, and E. Becirovic. Integrated power quality monitoring system and the benefits of integrating smart meters. In *Proc. 8th Intl. Conf. Compatibility and Power Electronics (CPE)*, pages 86–91, Jun. 2013.

[161] T. Baldwin, D. Kelle, J. Cordova, and N. Beneby. Fault locating in distribution networks with the aid of advanced metering infrastructure. In *Proc. Power Systems Conf. (PSC), Clemson University*, pages 1–8, Mar. 2014.

[162] T. Khalifa, K. Naik, and A. Nayak. A survey of communication protocols for automatic meter reading applications. *IEEE Trans. Commun.*, 13(2):168–182, Feb. 2011.

[163] M. Popa. Smart meters reading through power line communications. *Next Generation Info. Tech.*, 2(3):92–100, Aug. 2011.

[164] M. LeMay and C. A Gunter. Cumulative attestation kernels for embedded systems. *IEEE Trans. Smart Grid*, 3(2):744–760, Jun. 2012.

[165] K. Balachandran, R. L. Olsen, and J. M. Pedersen. Bandwidth analysis of smart meter network infrastructure. In *Proc. 16th Intl. Conf. Advanced Communication Technology (ICACT)*, pages 928–933, Feb. 2014.

[166] H. Saari, P. Koponen, E. Tahvanainen, and T. Lindholm. Remote reading and data management system for kWh-meters with power quality monitoring. In *Proc. 8th Intl. Conf. Metering and Tariffs for Energy Supply*, pages 11–15, Jul. 1996.

[167] D. Matheson, C. Jing, and F. Monforte. Meter data management for the electricity market. In *Proc. Intl. Conf. Probabilistic Methods Applied to Power Systems*, pages 118–122, Sept. 2004.

[168] USAID. Improving Power Distribution Company Operations to Accelerate Power Sector Reform. https://pdf.usaid.gov/pdf_docs/PNADJ549.pdf., Mar. 2005.

[169] Z. L. Gaing, C. N. Lu, and Y. T. Lin. Object-oriented design of trouble call analysis system on personal computer. *IEE Proceedings - Generation, Transmission and Distribution*, 142(2):167–172, Mar. 1995.

[170] X. Gu, H.-C. Wang, and J. Chen. Application of rough set-based distribution network fault location approach in trouble call management system. In *China Intl. Conf. on Electricity Distribution*, pages 1–5, Sept. 2012.

[171] C. N. Lu, M. T. Tsay, Y. J. Hwang, and Y. C. Lin. An artificial neural network based trouble call analysis. *IEEE Transactions on Power Delivery*, 9(3):1663–1668, Jul. 1994.

[172] R. Fischer, A. Laakonen, and N. Schulz. A general polling algorithm using a wireless AMR system for restoration confirmation. *IEEE Power Engineering Review*, 21(4):312–316, Apr. 2001.

[173] B. Amini, S. H. Khatoonabadi, and A. Zamanifar. Trouble call based outage determination in power distribution networks using anfis. In *Intl. Conf. on Power Electronics and Drives Systems*, volume 2, pages 1634–1639, Nov. 2005.

[174] C. S. Chang and F. S. Wen. Tabu search based approach to trouble call analysis [in lv power distribution]. *IEE Proceedings - Generation, Transmission and Distribution*, 145(6):731–738, Nov. 1998.

[175] H.-J. Chuang, C.-H. Lin, C.-S. Chen, C.-C. Yun, C.-Y. Ho, and C.-S. Li. Design of a knowledge based trouble call system with colored petri net models. In *IEEE/PES Transmission Distribution Conf. Exposition: Asia and Pacific*, pages 1–6, Aug. 2005.

[176] M. T. Tsay, W. M. Lin, and A. J. Hwang. A reliability model based trouble call analysis involving old secondary circuits. In *Intl. Conf. on Power System Technology*, volume 1, pages 270–274, Aug. 1998.

[177] S. Qiu, L. Guan, and H. Wang. An improved distribution network reliability analysis algorithm considering complex load transfer and scheduled outage. In *Intl. Conf. on Power System Technology*, pages 1–7, Oct. 2010.

[178] Y. Wang, C. Liu, M. Shahidehpour, and C. Guo. Critical components for maintenance outage scheduling considering weather conditions and common mode outages in reconfigurable distribution systems. *IEEE Transactions on Smart Grid*, 7(6):2807–2816, Nov. 2016.

[179] J. S. Wu, T. E. Lee, C. T. Tsai, T. H. Chang, and S. H. Tsai. A fuzzy rule-based system for crew management of distribution systems in large-scale multiple outages. In *Intl. Conf. on Power System Technology*, volume 2, pages 1084–1089, Nov. 2004.

[180] E. M. Morgado and J. P. Martins. An AI-based approach to crew scheduling. In *Proceedings of 9th IEEE Conf. on Artificial Intelligence for Applications*, pages 71–77, Mar. 1993.

[181] ENSTO. SAIDI and SAIFI indices guiding towards more reliable distribution network. https://www.ensto.com/company/newsroom/articles/saidi\-and\-saifi\-indices\-guiding\-towards\-more\-reliable\-distribution\-network/., 2016.

[182] J. S. Wu, T. E. Lee, and K. S. Lo. An evolutionary-based approach for outage scheduling of electrical distribution feeders. In *First Intl. Conf. on Innovative Computing, Information and Control - Volume I (ICICIC'06)*, volume 1, pages 14–17, Aug. 2006.

[183] J. S. Wu, T. E. Lee, C. Lee, C. P. Syu, and S. D. Su. An improved ga approach for distribution system outage and crew scheduling with google maps integration. In *Intl. Conf. on Machine Learning and Cybernetics*, volume 3, pages 967–973, Jul. 2011.

[184] J. S. Wu, T. E. Lee, and C. H. Cao. Intelligent crew and outage scheduling in electrical distribution system by hybrid generic algorithm. In *4th IEEE Conf. on Industrial Electronics and Applications*, pages 96–101, May 2009.

[185] Y. Tang, C. W. Ten, and L. E. Brown. Switching reconfiguration of fraud detection within an electrical distribution network. In *Resilience Week (RWS)*, pages 206–212, Sept. 2017.

Index

Printed in the United States
by Baker & Taylor Publisher Services